£5.95

An Eye on the Environment

an art education project

Joicey

edited by Olivia Bennett

Bell & Hyman

First published in 1986 by
BELL & HYMAN LIMITED
Denmark House
37–39 Queen Elizabeth Street
London SE1 2QB

British Library Cataloguing in Publication Data

Joicey, H. B.
An eye on the environment: an art and education project.
1. Human ecology
I. Title   III. Bennett, Olivia
333.7'024372   GF41

ISBN 0-7135-2622-X

Designed by Geoffrey Wadsley
Typeset in Great Britain by Cambridge Photosetting Services
Printed in Great Britain by Purnell Book Production Limited
Member of the BPCC Group

Acknowledgements

The author and publishers would like to thank all the teachers and pupils whose work is reproduced in this book, including those whose work was not originally submitted to the WWF project. Particular thanks are due to many Humberside teachers for their essential contributions. Every attempt has been made to acknowledge the author or artist of each item of work; however, we have not always been successful and would like to hear from anyone whose work has been reproduced but not credited.

We would also like to thank the following for permission to reproduce pictures and extracts from published works:
The Tate Gallery, London – 'Entrance to a Lane' by Graham Sutherland (photograph: John Webb)
Ronald Blythe, *Akenfield*, Penguin Books Ltd
Charles Chaplin, *My Autobiography*, Penguin Books Ltd
Keith Vaughan, *Journals and Drawings*, Alan Ross, London, 1966

Contents

The World Wildlife Fund

The World Wildlife Fund is an international conservation foundation with national groups around the world, including Britain. It was launched in 1961 to raise money for the conservation of nature and natural resources. So far it has channelled over one hundred million pounds into projects in 135 different countries.

Through the funding of national parks, nature reserves, environmental education provision and anti-poaching groups, and through the promotion of conservation legislation and environmentally sensitive agricultural and industrial practices, the World Wildlife Fund has protected individual species of animals and plants, as well as entire habitats such as rain forests, marshes, islands, meadow-land and coastal areas throughout the world.

WWF-UK is committed to an education programme that aims to produce resource materials that enable teachers to bring environmental issues into everyday classroom teaching at all levels of education. The materials are designed to give young people knowledge and experience that allows them to make informed personal judgements about environmental issues. Resources are being developed for most subjects of the school curriculum, making use of the inherent qualities of each subject to develop specific aspects of environmental understanding and sensitivity. These resources are produced from direct classroom work, from the work of groups of teachers, or by subject experts in conjunction with classroom trials.

The content of this book was generated by a school-based project that invited teachers and pupils to use art activity to develop their environmental sensitivity. The project was organised by the World Wildlife Fund with the Association of Art Advisers and sponsored by the Colonial Mutual Assurance Society Ltd.

Preface

To enjoy lying in the sun and watching the changing patterns of cloud and sky requires no action other than switching on our senses. Enjoyment of the environment and an awareness of how it affects the way we feel is the common ground of environmental care. This book is about how young people can be encouraged to switch on their senses and be drawn into a personal relationship with the environment which will enrich their lives and also benefit the environment.

Peter Scott.

Sir Peter Scott

Introduction

We are all guilty of looking without seeing. Children on the way to school pass by a wealth of visual detail and events, most of which they simply do not notice. We rush off to work and in our concentration on the deadline or meeting which looms ahead, we fail to see what is going on around us. Even when we are at leisure, we bury ourselves in novels, films and television programmes about the loves and lives of others, while all around us – in our homes and our communities — real dramas unfold without us even realising it.

The scale of these dramas may, of course, be small, particularly compared to those the mass media offer us. Some years ago, my attention was drawn to a passage in *Akenfield* by Ronald Blythe which is about, among other things, this question of scale.

> **The old village people communed with nature but the youngsters don't do this either. The old people think deeply.**
>
> **They are great observers. They will walk and see everything. They didn't move far so their eyes are trained to see the fine detail of a small place. They'll say, 'The beans are a bit higher on the stalk this year. . . .' I help to run the school farm but I'd never notice things like that. The old men can describe exactly how the ploughing turns over in a particular field. They recognize a beauty and it is this which they really worship. Not with words — with their eyes. Will these boys be like this when they are old? I'm just not sure. Nobody is trying to bring it out in them. Nobody says to them, 'This is heritage.' Somebody should be saying to them, 'Let's go and *look*. . . .'**

Ronald Blyth, *Akenfield,*
Penguin

This passage impressed me as conveying a fine, simple aim which applies to much that we are seeking to achieve in education. We should be encouraging children to look, to see and to know; and training them 'to see the fine detail of a small place'.

This approach to education coincides with a number of initiatives which actively encourage the child and teacher to look at their immediate environment, and increase their understanding of it: the Schools Council History Project, Schools Council Science 5–13 Project, developments in English associated with personal writing, and the *Art and the built environment* project, are a few obvious examples.

This view, however, seems to run counter to a general and growing tendency to ignore the 'small place'. Today's emphasis is on speed: there's no time to observe the 'fine detail'. Keith Vaughan, the painter, noted signs of this in his diary:

> **22 January 1944 Further evidence of the way things are going was offered by the MOI film of air service after the war. Here is the smooth totalitarian efficiency of the large air port. The control tower with charts and panels and clipped conversation. The flare path, the runways, and the liners arriving to the minute from all corners of the earth. The poker-faced pilots with their brand new clothes and old jokes. Breakfast in Leningrad, lunch in Le Bourget, and on in time for the evening show in Madison Square Garden. This, we are confidently asked to believe, is the answer to world organization. Getting about. 'Because,' as one earnest, baby-faced American pilot puts it, 'it seems to me that when folks get to know more about each other and understand each other better wars just won't happen.' Maybe. But has it occurred to anyone to wonder how much one will get to know the Russian people through a succession of breakfasts in Leningrad, or how much experience can be gained at the pilots' bar at San Francisco or Los Angeles.**

and he concluded in his entry for that day,

> **It is as well to know where one is going before one leaves one place for another. It is essential to know if one travels fast. Travelling slowly one comes upon interesting things by the way which were not foreseen. So that even if one arrives finally at the wrong place the journey was not wasted. But the greater the speed the more accurate must the aim be. Fast travel eliminates everything that lies by the way. There is only the beginning and the end — least satisfactory of positions. Only the very sure should travel fast. Our dilemma is that the speed and facility of travel increases in inverse ratio to our sense of direction.**

Keith Vaughan,
Journals and Drawings,
Alan Ross

In the educational process, there is much evidence to show that we learn most from our experience with 'everything that lies by the way'.

This emphasis on fast travel, in which a journey is merely the time it takes to get from A to B and not an experience in its own right, is alienating. It 'eliminates everything that lies by the way'. We lose our sense of direction, in more ways than one. By adopting a different view, one which values the journey itself and the fine detail it can reveal, we can become more in touch with our environment. We see things we would otherwise miss and so we can develop an intimate relationship with the surrounding landscape. Thus we develop a skill and a sensitivity which, once learned, is likely to stay with us for ever. It can have a profound effect on the way we view the world we live in.

This is why using art as a means to open our eyes to the environment is so relevant to the World Wildlife Fund's educational programme, which calls for changes in our attitude and behaviour so that people may live in harmony with the natural world. These changes involve learning to see the fine detail and becoming aware of 'everything that lies by the way'. Undoubtedly, the process of drawing and painting is an excellent means of helping children in that educational process.

Drawing a closely observed plant will provide the artist with a sound knowledge of the shape and form of that plant. Drawing a landscape will give the artist an intimate knowledge and understanding of the different elements which constitute that landscape. Drawing a building gives the artist an insight into architectural form, whilst sitting and drawing in a street allows the artist to hear and feel the life of that street.

The same applies to the child who is given the chance to become involved in the process of drawing, painting and three-dimensional work, for it contains two vital ingredients for educational development. It encourages the child to look hard at the subject and it gives them time to reflect. Thus it encourages the experience of seeing and knowing. This is true for everyone, whatever their age or level of skill. The difficulty is being sufficiently convinced that this is so; and then knowing how to start. Once that first move has been made, however, the value of this way of working soon reveals itself. The process of looking and drawing causes people to reflect not only upon the subject but also upon their response to it. There is time to absorb the 'fine detail'.

An Eye on the Environment seeks to provide teachers with examples of ways in which they may help children to experience this process. It also shows how some teachers and pupils have responded to this way of working. It incorporates teachers' reports and examples of work from a number of schools, some of which took part in the World Wildlife Fund's project called 'Looks Natural'. The following extract from a teacher's report so exemplifies this approach that I have reproduced it at some length. It demonstrates admirably just how possible it is to offer a large group of children a variety of ways of increasing their understanding of the 'fine detail of a small place'.

I wanted the children to see their immediate surroundings with a new awareness. The site I chose was forbidden ground for the children, so naturally they loved going on it! We work in a small modern school, next to an enormous Victorian

Gothic Church. A corner of the Church grounds overlooks our playground, but is separated from us by two retaining walls. It is dark, gloomy, full of weeds and rubbish; the walls of the Church are covered with bird droppings, there are bits of wire and old bricks, and it is also quite high above the ground. Great!

Their previous work in the First School had mainly been large-scale painting on sugar paper, using colours already mixed for them. I wanted to experiment with, and gain confidence in, new materials, to develop new skills, and acquire an increasing awareness of colour and texture.

Most of all, I wanted the children to have fun. It is no good providing eight year olds with new experiences, new materials, if there is no enjoyment or freedom or excitement. I also wanted each child to have something they could keep, so everyone made their own folder-book. This was to be a record of their experiences which I grouped into 6 main activities. Each activity was separate, yet led into the next, and built up into a fairly detailed study of this piece of ground. There was to be much discussion, and I wanted Art to link up with other subjects: in R.E. we went from Genesis via David and the Psalms to N.T. Parables!

In Maths we covered area and graph work based on the site. Studies of grass led to work on the major cereals of the world and so on.

The children cut pictures from magazines, and parents bought nature books for them and most children became much more interested in their own back gardens, and brought objects to school daily.

We went out, looked, came back and tried to recollect what we had seen — retaining walls of old crumbling bricks, weed- and rubbish-strewn waste ground, the dark Church behind. The pictures were drawn in wax and pencil crayons, on white cartridge and coloured with an ink wash. All of them loved the excitement of painting over the crayons with the inks. Then they made a list of what they had seen on the ground. Mainly rubbish was noted, as you will see from Keilly's list. We then had a long discussion, with many questions being asked. Why were there bird droppings on the wall? Could we see their nests? What sort were they? What do birds eat? What do insects and small creatures eat? What plants and weeds grew there? We decided we had better go back and have another, much closer look!

This time, on a lovely sunny Autumn day, we knelt, sniffed, touched, squashed, rubbed and examined minutely everything we saw. We found ladybirds, leaves, stones, twigs, feathers and parts of dead birds, spiders, webs, bees, weeds, seeds, shells (at one time the whole area was under the sea). We listened to the birds, the church bells, the wind, distant traffic, and enjoyed the sun. We came in very dirty, very happy, very excited, only to be told that a mild week-killer had been applied to the ground the night before. Have you ever scrubbed 29 eight year olds in a school sink? Eventually, they mounted their finds on individual sheets of paper, and then wrote their reasons for choosing them. The children then drew, pressed, rubbed, printed and painted different objects,

James Knight
Aged 8
St. Mary's First and
Middle School
Grimsby

Keilly Dunks
Aged 8
St. Mary's First and
Middle School
Grimsby

giving great attention to colour and texture. We used a magnifying glass to examine objects closely, and this led to individual work on insects, birds and plants.

The children were now realising the waste ground was more than just a collection of rubbish but the home of insects and small creatures. What must it seem like to them? Using plasticine and boards, the children built up their own three-dimensional landscapes, using natural materials they had collected. There was a great deal of competition over this, and no end of a mess. When they were satisfied, they drew their landscapes, in oil pastels, on cartridge paper.

The children now began to study the birds and animals found in the area. Most chose birds: sea-gulls from the coast, owls who live in the nearby belfry, pigeons who live all over. They had nature books and posters, to study and to give them knowledge, but drew their pictures from memory. I gave them free choice of materials and paper, everyone chose oil pastels and white cartridge. They spent far more time on this than I had expected, using dinner and playtimes to do extra work.

I'll be honest, I wanted an interesting gimmick here. I wanted children to see the relationship between creatures and their surrounding — a sort of 'Lift up and Look' project. I planned three layers, and gave the children one example, from life. I lifted up a stone to show leaves clinging to the underside, with beetles underneath the leaves. Most children, inevitably, chose stones or leaves, but a few chose birds. All of them painted their work, and mixed their own colours.

I now felt the children needed a complete change — there is a limit to the concentration and interest of eight year olds, no matter how important the subject. Since we had read a lot of poems about nature, I let them choose, decorate and illustrate the poem they had liked best. They used felt tips, on white cartridge.

What had we achieved? The children had all realised the need to preserve and care for wild life and nature — on a much larger scale than we had dealt with — and were quite vehement in their denunciation of pollution and destruction. I was amazed at the rapidity of the change from 'what a lot of rubbish' to heated discussions of the usefulness of worms as opposed to the benefits provided by ladybirds; from stepping on weeds and stones to carefully lifting them up and examining them closely. They had also (I hope) seen that life, excitement, colour and interest are everywhere, even in a derelict corner of an old churchyard.

P.S. *I'd* had fun.

Chris Chappel
St. Mary's First and Middle School
Grimsby

A Strategy for Seeing and Understanding

It is all very well to write of the value of drawing and say that children should have more opportunities to become involved in it. The fact is that, although drawing is an easy act to become physically engaged with, and reasonable results are not difficult to achieve, few teachers or pupils perceive it this way. Confronted with a blank piece of paper on which they must make a mark, the confidence of many pupils rapidly disappears. They hold in their minds an image of what the drawing should be like, and thus their first tentative efforts are too often an anti-climax which, they feel, merely expose their lack of ability.

Similarly, pupils sent out to gather 'rubbings' or 'textures', or to make drawings from the landscape/playground discover that there is so much to see that often they simply don't know where to start. When they do put pencil to paper, they don't know how many sketches to make or what size they should be. So they tentatively put their bits together and present untidy, loose and ugly pieces of work which have the stamp of failure upon them.

The way to overcome these stumbling blocks is to develop a strategy which enables those first marks on the paper to succeed and which gives such exercises a real sense of purpose.

Starting points

One effective way of providing the necessary direction is to ask children to gather specific information on specially prepared drawing sheets. They also need to be sent out with explicit instructions as to what they should be looking at: branches, railings, clouds and so on. This should avoid the problem of apparent failure of execution.

The environment of the school is a wonderful source of inspirational material to develop the children's ability to look without preconception. I usually give the children a visual theme to look for such as 'lines' — objects made of lines or objects with lines on them — or the theme could be shape, tone, colour, form, space, texture, pattern, relationships, or harmony, developing finally with the oldest children to such themes as atmosphere of weather or mood of place. This gives them a starting point.

L. A. Dear
Dulwich College Prep. School, Kent

There is nothing magical or mysterious about drawing sheets. They simply allow a child to maintain a greater semblance of order and control in their work. The completed sheet, whatever the level of skill, does present a good visual image. Above all, it gives some sense of achievement rather than immediately exposing weaknesses.

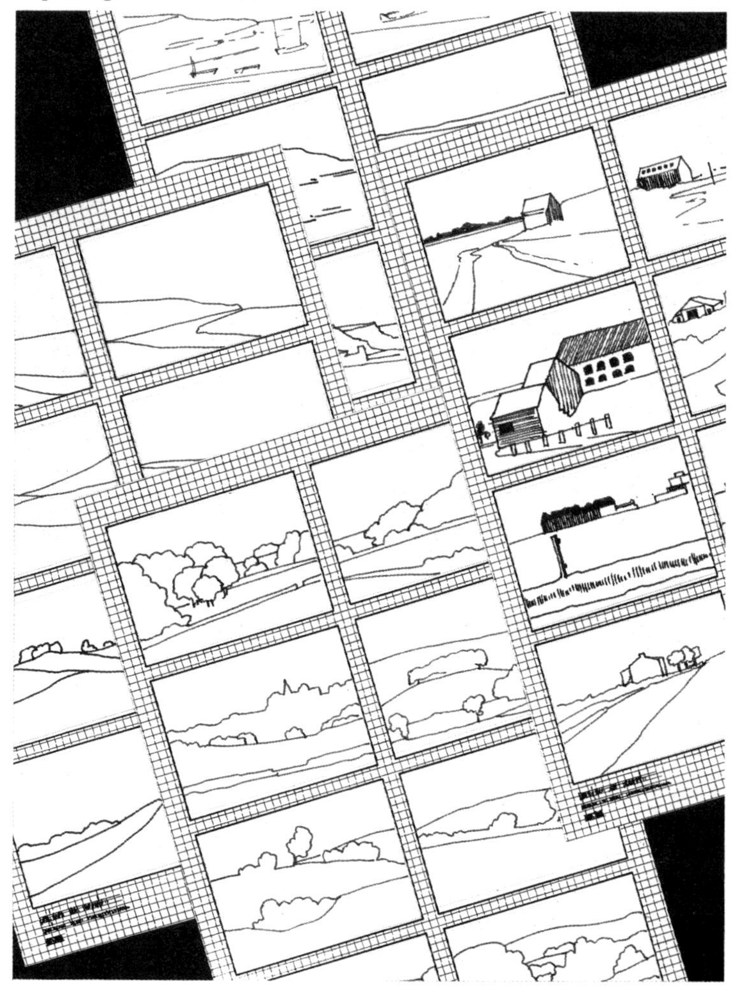

The boxes in the worksheets can be designed to fit whatever resources the school can make use of.

I have a small collection of unusual insects brought back from overseas trips. These always interest the children but I then insist that they bring in their own resource material from the 40 acres of mature parkland we are lucky enough to have as our school grounds.

Wendy Nicholson
Gillott's School, Henley-on-Thames

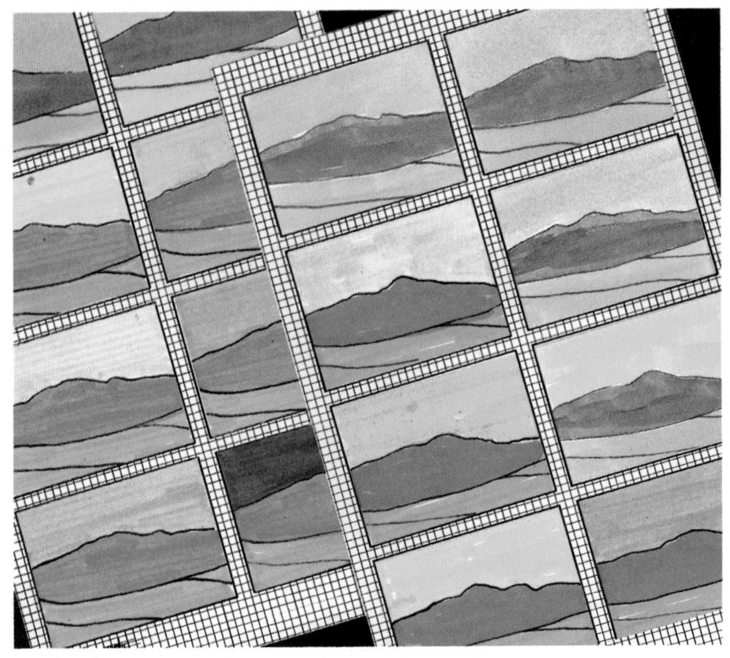

They can even be used to look at colours.

Our children are bombarded with instant art, influenced constantly by cartoons, spoiled by the availability of a myriad of felt tips — all exciting in their own way, but certain skills are in danger of being lost for good, including the ability to discover colours and the urge to investigate and care for the environment.

Wendy Nicholson
Gillott's School, Henley on Thames

What we are considering here is not so much an approach to art, but rather a means of gathering information which applies to all subjects in the curriculum. The use of worksheets like those shown can provide a means of directing the pupils to gather specific visual information appropriate to whatever project may be underway.

Rural Studies makes considerable use of the fine arts as a strategy for stimulating interest in the environment. The subject is offered as an option.

As various topics are completed, e.g. 'The Weather', 'Fungi', 'Fruits' etc., the pupils are encouraged to improve the construction and textures of their mainstream art by incorporating the knowledge gained in Rural Studies.

Art is employed continuously to focus attention onto detail of composition of living things; to cultivate an appreciation of the aesthetic splendour of the area; and to create an awareness of the desperate need to preserve the irreplaceable treasures that form part of our dwindling heritage.

From concrete to abstract

Thus, in order to encourage children to look, gather and use information, there must be a clear strategy for seeing and understanding. This book presents a two-fold strategy. It suggests themes, or 'areas of interest' which are likely to prove a rich source of stimuli. It then provides a subsidiary series of 'elements', to help children explore these themes with real direction and intensity. These elements (line, point, texture, rhythm, form, colour, tone, pattern and shape) have long been established by art educationalists as significant in any analysis of visual images. All too often, however, teachers start with these abstract concepts and then move to more concrete ideas. The strategy employed in this book is to reverse this process. Thus Part I looks at distinct and concrete areas of interest such as the sky, fields and spaces, buildings in a landscape, and so on. Part II explores the abstract concepts, which can be then applied to the themes in Part I in order to provide a direction and a structure for looking and for gathering information.

It is hoped that this approach will make the process of looking easier and therefore more successful. The aim is to help the teacher and the child find ways of increasing their understanding of the environment, while at the same time developing their painting and drawing skills and gaining a better grasp of visual education. But it should never be forgotten that such work should be fun. Take account of the 'capricious, mobile and unpredictable mind': the strategy is not meant to contain the experience but rather to encourage it to develop.

I was overwhelmed by the large sweeping views of the coast, the sea and the sky, stimulated by the texture of the moors, the shoreline and the cliffs, rather than the small detail. I found it most difficult to concentrate on the objects found beside the derelict farm buildings and hedges in the area. Photography seemed the best way to capture these objects and their surroundings on the day that I was working, and no doubt, would prove a useful resource in the future. . . . My attention turned toward horizons and divisions across the landscape which made shapes, irregular shapes on the horizon and between the fields. Textural qualities captured my interest together with the autumnal colours which crept through the fauna. . . . When I found an area which I could enjoy, I felt frustrated by the lack of time though it was possible to make small sketches. . . . This is a beautiful time of the year in which to be in the landscape. I wanted to explore the richness of the colour to be seen, particularly on the moors. There was the spectacular contrast between the reds of the moorland and the greens of the hills and valleys.

Nancy Carter
Vale of Ancholme School, Brigg

Putting the Strategy into Practice

One of the effective qualities of art and design as a learning vehicle is that it places emphasis on the child learning for his or herself. The rudimentary skills of drawing and painting are comparatively easy to learn. From then on it can be a matter of children largely controlling their own development and learning. When a child draws the skull of a sheep, after the first mark has been made, it is the skull which tells the child whether the drawing is succeeding or not. He or she can continually assess their progress by comparing their drawing with the reality of the skull. This is a prime value of all still life and objective drawing. The object is there to teach whoever is drawing.

Indeed, it provides the circumstances for learning a good deal more than how to symbolise reality on a piece of paper. Drawing skulls, bones, plant formations, land and town scapes is an excellent means of acquiring a whole range of knowledge about the subject. The direction of learning will take its own course. In the case of the skull, for example, the child may become aware of and take interest in bone structure or in simple patterns of line; it may fire morbid curiosities in the child; or remind him or her of other organic forms found in the landscape.

The keys to success with this approach lie with the object itself, and the attitude of the pupil. Firstly, the object must excite the curiosity and observational powers of the child. If it succeeds, then the child's attitude will be positive and enthusiastic. Secondly, the child needs to be prepared and confident about making that crucial first mark. It is true that their degree of skill affects this initial move, but any inhibitions gradually disappear as the object takes over as teacher and the child becomes absorbed in the process of drawing.

Making that first mark is a matter of confidence. Knowledge of technique and skill help a great deal, of course, but let us assume that a large number of children lack, or feel they lack those. Another major confidence booster is enthusiasm. It can overcome many fears, doubts and uncertainties. If the object fires the pupils' enthusiasm, then the problem of making that first mark is largely resolved.

Thus the source material is crucial to this strategy. It must excite and arouse the interest of the children. It must also provide scope for a range of learning to take place, for the value of taking this approach to art and design education is that it can provide teachers with a strategy for learning in general. A well constructed and controlled drawing lesson offers the teacher an environment in which children are actively directing their own learning on a variety of levels and in a variety of directions. Children asked to draw a sod of turf may find themselves improving their drawing skills, gathering some knowledge of plant structure and root growth, and perhaps gaining a perception of the world in which small creatures live, which could then relate to their current storytelling and reading work.

All this ties in directly with the need for schools to offer a broader context in which children can develop their interests. This is why selection of the source material is so important: a carefully considered choice should ensure that there is plenty of educational scope, but, above all, that the children, actively directing their own learning, find themselves involved with areas of interest which coincide with the teacher's knowledge and the aims of the syllabus.

An art education project related to looking at the environment is an obvious example of that kind of strategy. Drawing in the environment provides immediate links with the humanities, the sciences and with poetry, music and literature. The only problem is that the overall concept of this can seem so promising that the 'fine detail' gets overlooked. Thus the principle of making sure that those asked to go out and produce drawings have their attention drawn to specific detail has to be applied throughout the project. A class will need a tight structure in order to make sense of the vast amount of information which can constitute an environment.

Yet a tight structure need not and should not be confining. Children asked to look at the pattern in the bark of a tree will gradually grow aware both of the tree and of the detail within the bark. They will start to take in the tree in its surroundings; whether it stands alone or is part of a copse or wood. They'll notice leaves and seeds on the ground; birds in the tree; insects on its surface; the shape of its roots etc. But this realisation and appreciation of the wider detail will come as a result of their eyes being directed towards the fine detail of the bark.

The tools of the strategy

The project offers the teacher ten 'areas of interest'. These were selected for the following reasons: they have proved to be areas capable of producing enthusiasm and interest on the part of both teacher and pupil; they provide plenty of scope for ideas and experience to develop, which relate to other areas of the

This is the story of a Dandelion seed.

I was sitting there watching the hedges sway this way and then that way. It was a fine day and all the bees were at work. My bright yellow petals sway, and then I felt something, but it was a bee thats all.

Summer was almost over. I was getting old now. My bright yellow petals were no more yellow, many were brown. Soon I will be blown away by the wind and I will be separated from my friends. I will miss them very much.

Next day I was just feeling happier but it was too late, I had a white circle on me. I was very old.

Then a couple of my friends had been blown away by the wind and by now I was feeling very very frightened indeed. I tried to cling on but it was no use, the wind had blown me away. I was the last one.

By now I was high up in the air. I was feeling very lonely. Then I saw my best friend. We joined together. Then I was blown down onto the ground. I got very comfortable.

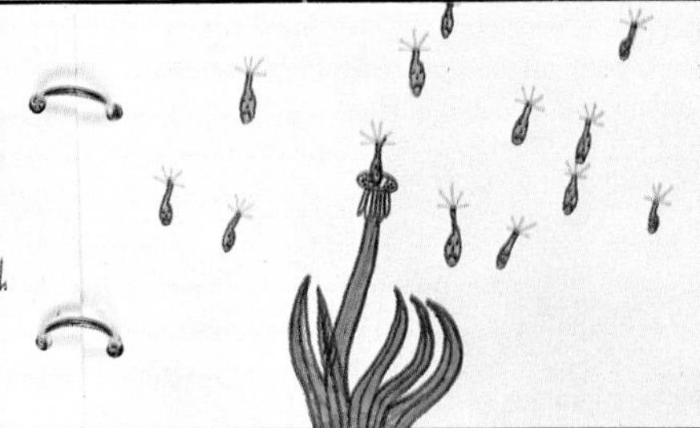

It started to rain. I was all wet and soggy. Then it started to thunder and lighten.

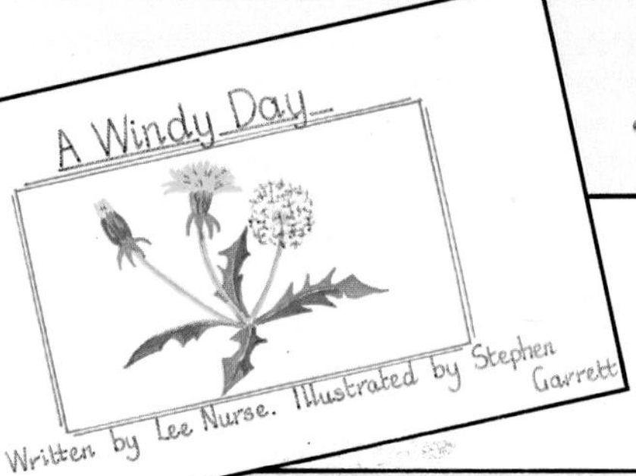

Then I was sound asleep.

When I woke up I was on a rock. I could not see very well because I was in the mud and it was almost summer again.

I was my normal self again. I was going to make new friends once more and I could smell fresh air again and hear the hum of bees.

. . . .

Windy Day tells the ry of a dandelion ed and is an example the kind of work that n be produced by opting the strategies tlined in this book.

e Nurse and Stephen rrett
ckingham Junior and ant School
therham

curriculum; and they provide ways of encouraging the children to refine their skills of not only painting, drawing and three-dimensional work, but also seeing and understanding.

The project then explores nine 'elements' which have been commonly accepted as significant in any analysis of visual and tactile form. Although the 'areas of interest' and the 'elements' are explored separately, they are complementary and should be used together. Familiarity with and understanding of the elements explained in Part II are essential to the activities suggested at the end of each area of interest. Non-specialist teachers may therefore wish to complete Part II before tackling the activities. The areas of interest highlight only a general direction for looking; the elements provide a more specific way of exploring a particular aspect of the environment. They provide the tools, or language, of analysis. And once children have a language in which to articulate what they are seeing and doing, their understanding of it increases. The experience is more likely to become one which can be built upon.

Equally important, the structure of the project takes account of the abstract nature of the elements and the difficulty children have in conceptualising them. Many attempts to teach using these elements have failed, largely because children have great difficulty in understanding what they mean. Exercises which consist of making 'points' or 'lines' in different shaped rectangles are often difficult for children to relate to. They can't 'see' how it divides up the space or alters the point of focus. But drawing a man (or a dog or a tractor) in a field, and then moving it to three different places, makes more sense and helps the child to appreciate how the single figure or 'point' does become a focal point and does alter their perception of the overall space. Similarly, an exercise where the 'line' is a horizon in a known environment is likely to mean more.

Thus the areas of interest provide the concrete examples through which the pupil can come to understand more abstract principles. At the same time, these abstract principles provide a greater degree of direction and focus to the areas of interest. Looking at 'fields and spaces' or 'buildings in a landscape' can be a daunting experience. Where does the child start? But ask him or her to take special note of the 'texture' or 'pattern' in those fields or buildings, and you have given them a structure for gathering information and the means to start developing a language for analysing and explaining what it is they are seeing, and trying to express through their drawing. In other words, you give them the means to start developing a greater awareness of and relationship with their environment. And that is the aim of **An Eye on the Environment.**

Activities and Materials

This book includes a series of suggestions and ideas, the aim of which is to stimulate further ways of working within and from the environment. They are not exclusive and assume no right or wrong way of working, stressing only that the selected approach should seek always to excite and interest, to underline existing strengths and to encourage children to succeed. They do assume, however, that there is such a thing as bad artwork, and that is when the finished work lacks creative impulse and contains nothing from the individual who has produced it.

The activities come in Part I, at the end of each area of interest, but they draw upon familiarity with and understanding of the elements described in Part II.

Media

Any medium which children can use to record and develop what they see is suitable. A long list of the various graphic media is unnecessary but the following comments should give some idea of general directions.

Drawing

The text contains many references to drawing as a means of quickly recording what is being observed. However, drawing can become an extended activity in itself and time should sometimes be given to individual drawings. It is a medium which can be used to develop ideas and make notes and recordings to furnish the imaginative and sensory experience which art should be.

Painting

Drawing implies the use of one colour and perhaps reliance on the extensive use of line. Painting has few such inherent inhibitions. The range of available materials is vast, yet too often schools depend upon one type of paint, one method of painting. It is important in all creative media to develop a confidence in materials but once that has been achieved, the pupils' experience should gradually be expanded.

Printmaking

Printmaking can be used to record what is being seen, but it is more appropriate as a means of recreating or developing work and ideas already recorded in drawings. Different types of printmaking may be more applicable to particular interpretations, for example, screen printing to colour; potato, card and texture to pattern and tone; etching to line work; and monoprinting to the whole range.

Three-dimensional work

Media associated with sculpture and three-dimensional work are most often used to realise imaginative ideas or extend work originally produced as drawings. A better understanding of these media might be achieved if they were used in observational exercises, for example, clay for an observed 'still life', figure or animal; and card, scrap, glue and wire to reproduce observed objects, buildings or landscapes.

Fabric and Textiles

It is possible to 'draw from life' using embroidery techniques but fabrics and textiles will be generally more applicable to the developmental stage of work. Screen printing can be used on fabrics, while patchwork and machine embroidery can be used to recreate landscapes and environmental settings quite easily, and with highly satisfactory results. The book's areas of interest, with their scope for pastoral work, natural settings and colour harmonies, make fabric work full of potential.

Photography, Film and Video

The desirability of recording as much information as possible in the field makes photography a useful ally and learning to use the viewfinder to isolate points and moments of interest can be good training. However, the process of drawing enables the child to develop their knowledge of the subject in a way photography does not and, given the objectives of this project, this should always be remembered. Use of the medium should build on the equipment available. For example, simple polaroids can be used to record movements in cloud formations over a given period, or changes in the effect of light on a building or landscape, as well as seasonal changes. The more sophisticated the equipment, the greater the possibilities.

Worksheets

Examples are given of a type of worksheet found to be suitable for the information-gathering exercises encouraged throughout this book. They are intended only as a guide, and it is suggested that teachers draw up worksheets appropriate to their own teaching.

General activities

The following general activities could be applied to a variety of environmental projects.

1 Ask children to recall a specific environment that they know and like. They can:
 (i) Draw a map from memory.
 (ii) Make a drawing from memory.
 (iii) List things which they remember about the place.

2 Ask children to describe a garden area they know:
 (i) Make a map or plan or diagram of the garden.
 (ii) Make a drawing of the garden from memory.
 (iii) Complete a short questionnaire about what can be found there:
  (a) types of plants
  (b) trees
  (c) animals and insects
  (d) birds
  (e) types of stone
  (f) colours
  (g) special features.

3 Explore a specific environment, for example a local park, and
 (i) Using various drawing techniques, make a visual record of:
  (a) plants
  (b) trees
  (c) shrubs
  (d) walls
  (e) features
  (f) rockeries
  (g) ponds and lakes
  (h) furniture
  (i) animals and insects
  (j) people
  (k) buildings.
 (ii) Keep a visual record and nature diary of the environment.
 (iii) Make a visual analysis of the area — shape and form, textures, colours, patterns, etc.
 (iv) Using the information gathered, re-create what has been seen in terms of paintings, prints, models and fabric work.
 (v) Stage a follow-up exhibition, for example, it might be possible to hold a small exhibition in a building within the local park.

4 Ask children to identify a 'personal environment' and to keep a visual diary of that place, using sketchbooks, folders and/or ring binders and plastic wallets.

5 Study birds, animals and insects in a specified area: the school, a local space, or even a selected area some distance away. Make drawings of the 'creatures', using models, and their habitats, and find out their history and any stories connected with them.

6 Identify areas of visual interest accessible to the pupils and prepare suitable sketchbooks, records and worksheets. These should include follow-up suggestions for which the sketches and notes can be used. Suitable sites include:
 (i) the local tip
 (ii) football ground
 (iii) park or garden of a large house
 (iv) derelict buildings
 (v) railway station
 (vi) industrial site
 (vii) allotments
 (viii) the High Street
 (ix) a hill top.

7 With the children, build up a resource collection to help study the environment which includes:
 (i) photographs and illustrations from books and magazines
 (ii) model houses and buildings
 (iii) actual plants
 (iv) stuffed animals and preserved insects
 (v) paintings and drawings from children
 (vi) polaroid photographs.

8 Make use of regular field trips to specified environments such as a small-

holding, farm, particular landscape, or industrialised area.

9 Make a three-dimensional environment in school, e.g. a model of:
   (a) a small street
   (b) walls with posters and/or vegetation
   (c) garden(s)
   (d) a farm
   (e) fields.
   (i) Write a story about that environment.
   (ii) Keep a diary of the model area.
   (iii) Use the model as a subject for imaginative paintings and prints.

10 Maintain a school, class or personal visual diary on the subject of the school environment.

11 Organise a Day in the Field, using an established format so that each child, class and teacher share a similar experience in terms of making a visual response to an environment.

12 Collections:
   (i) Collect a given number of 'objects' on a walk, using, for example, a plastic tool box with compartments and a transparent cover. Specify beforehand the type of objects to be collected.
   (ii) In the same way, collect examples of stones, wood, shells, textures, etc.

13 Ask children to make drawings and colour sketches of particular fauna throughout the seasons.

14 Make a recorded journey through a carefully selected environment, visually noting what is seen at specified points.

A day in the field revealed this contrast between the natural fauna and the man-made object

Melanie Webster
Aged 11
Westwoodside C.E. School
Doncaster

Part I Areas of Interest

Roads, Railways, Tracks and Paths

One of the difficulties in drawing or painting the landscape is making sense of what is being looked at. A group of trees, a collection of buildings, various plants, a hill or some fields stand alone on the page, waiting to be put into some sort of context. Literary tradition is so strong that we look for some story, theme or pattern which links everything together and makes sense. A road or railway, a track or a path can do this. By running through a landscape, they give it a sense of purpose or order.

Two teachers asked to explore this theme on one afternoon in a moorland landscape came back with the following notes:

> ***Railway Lines:*** **Converging parallel lines — perspective — cuttings into the landscape — tunnels — earth bankings (earth sculpture): we became aware of the dramatic nature of the man-made environment and thought how the whole area was ripe for three-dimensional work. We thought of ceramics: the cutting into and the building on top of clay. We thought of prints with lino: the cutting into and building on top of the surface.**
>
> ***Tracks.*** **Although we didn't have the time to seek out what we understood to be tracks, we thought about their potential. Animal tracks, prints across the beach, through the snow. We thought about tracks in relation to texture — the changes in grass colour; we thought of them as something non-permanent and we thought of printmaking — wet footprints, the trail of a snail. Secret tracks and wondering who uses them?**
>
> ***Roads:*** **They were undulating. A road led down to and under a shallow river — a ford. The road was dissolved by the rushing water. The colour of the road above and below the water. The fluid texture of the road. We thought of the different types of road — minor, major, motorways, bridle paths. I noticed we both drew the same undulating road from a sense of inferiority — a fear of something different.**
>
> ***Paths:*** **These often bent around the features of the landscape and so they drew our attention to the shapes of the fields. Although I was studying the linear quality in the landscape, the texture of the fields began to interest me.**

R. Tonks, Headlands School, Bridlington and J. Johnstone, Welholme Middle School, Grimsby

The rich potential of this area as a source of stimulus for further work both in the landscape itself and back in the classroom can be seen from these personal responses. Their ideas roam from two-dimensional media to that of clay, from natural history to

ese sketches
strate how the tracks
ve been used as the
rting point for looking
he landscape

ohnston (Teacher)
elholme Middle
hool
msby

imaginative invention, encompassing a wide variety of curriculum work in between.

These two teachers produced such ideas as a result of actually working in the landscape. Their task was to draw but they brought back not only drawings, but also evidence of the thoughts which had accompanied and were provoked by the artistic process. It can be the same for children. All they need is a teacher who has the confidence to send them out to draw and make that experience exciting and stimulating.

The potential difficulty in this strategy, however, lies in the organisation of time and resources. The confines of the school day or timetable sometimes make it hard to take children outside and work with them, even when the school has ready access to grounds. Groups can be large, teaching time limited, and external pressures often keep staff and children inside their classrooms. Yet the two teachers' report shows how many ideas and starting points can come from just one well-organised day. Their sketches, thoughts and reflections will provide enough work for a number of further sessions within the school. Such work will require a supply of appropriate resources such as photographs, branches, twigs, leaves, models and a range of bric-a-brac which will sustain the original ideas.

Too much emphasis on first-hand experience and objective drawing can give a wrong impression. Ultimately art is about the organisation of our emotional, imaginative and sensory experience. Work which is the 'product of the inner landscapes of the mind' should be an outcome of such activity. The drawings, paintings and three-dimensional exercises should be feeding work which is concerned with drama, feelings, mood and sensation — the sensations experienced from being in the landscape. It is this process which enables one to develop an intimate awareness of the environment.

The sweep of the road illustrates the fall and rise of the landscape and frames the huddle of buildings against the hillside

Mathew Carr
Aged 15
St Augustine's R.C. High School
Blackburn
Lancs

The road leads the eye into the centre of this landscape, highlighting the distant church and the wood stretching across the scene. Perhaps it suggests a story.

Rachael Day
Aged 16
South Hunsley School
North Ferriby

The road as a line takes the eye through and into the landscape

Dominic Glover
Aged 11
Dulwich College Prep. School
Kent

Suggested activities

1 (a) Make a series of drawings of roads, tracks, railway lines or paths showing how they can serve as lines dividing the space through which they pass.
(b) Make a collection of drawings showing how the paths, tracks, lines or roads can lead the eye to or from a particular place and give it importance.
(c) Looking at a particular landscape, show how the space created between roads, for example, can become an interesting shape or series of shapes.
(d) Similarly, show how the undulations of roads, the twisted movement of paths and tracks, and the bending of a railway line can emphasise the form of an environment.
(e) The rise and fall of a road passing into the distance, the surface of a footpath, delicate footprints repeated in mud and the tightly packed ballast on a railway track all make distinctive patterns. Make a series of drawings showing these.
(f) The surfaces of all the four examples listed in (e) above have particular textures. Ask children to draw these different textures or to note the varying textures to be found along a particular road or path.
(g) Changes in the quality of light and the weather, as well as the use of a particular area, cause corresponding variations in tonal qualities. Ask children to make a visual record of such different tonal values, noting how the type of surface, its thickness and the nature of its use affect tone and colour.

2 Make a model of a road, railway, track or path using a combination of scrap material and bought models. Give it as much 'life' as possible, incorporating stories about the place, examples of the flora and fauna that can be found there, and maps, paintings, drawings and models which re-create incidents associated with it.

3 Select an appropriate path or track and keep a 'visual diary' of it through a year.

4 Take a section of road, path, railway or track and study the various constituent elements, e.g. fauna, man-made objects and textures.

5 With the aim of extending the children's response to an area beyond the purely visual, explore it through an imaginative theme such as:
(a) the railway filled with ghosts
(b) the overgrown and untended path
(c) the road disappears over the hill
(d) tracks in the snow.

Fields and Spaces

The expanse of fields in the foreground draws attention to the line of housing and the environmental furniture of the estate.

Jonathan Yeo
Aged 9
Willows Middle School
Grimsby

Open spaces, wet, windy and cold. Fingers tightly holding brush and pencil.

Jean Hockenhull,
Riddings Comprehensive School, Scunthorpe

Sometimes it is sufficient to ask people just to go out and look. The quality of the landscape, the weather, the atmosphere and the sensation of the moment combine to create a mood and impose a structure on the artist.

Fields fitting together like bodies folded together, the yellow, greens and browns united in tone, edges defined by the dark texture of hedges, edging them in.

The way of seeing has already begun to determine the way of drawing, though it is always possible that the difficulty of making those first marks on paper may interrupt the vision.

Initially, the idea of looking at fields and spaces may seem at odds with the needs of the artist. The popular conception of rural or urban subject matter usually involves busy corners of undergrowth, woodland settings, or buildings tightly huddled together. Open fields or wide expanses of parkland may seem to offer little scope until the attention is drawn to what the space may be doing to the surrounding fringes.

It may be that the space encourages the eye to concentrate

on the thin wisp of trees running along the top or the delicate facade of a building at the edge. It may be that the broad expanse of grass or concrete draws attention to the huddle of cows or the solitary figure. On the other hand, the group of trees and shrubs or other surrounds may make the space appear as a strong, flat dominating force.

Spaces hard etched between the trees, their shapes shown dark and precise against the flat greyness of the sky.

Again, if the detail of the fields and open spaces is to be accessible to the children, if they are to become aware of what constitutes the landscape, then they will probably need some help as to where to start and what to look at. It may be that some will already sense a direction for their work:

Rich autumnal browns and greens, bright splashes of gold and yellow and over all a wash of soft grey mist.

Most, however, will benefit from being asked to look in a specific way:

Seeing the landscape in terms of interlocking shapes, the smooth and rolling body of the land, wearing the coloured pattern of the fields.

This is the way one teacher worked, when asked to go into a broad, bare landscape and look at the shape of the fields and moorland around her:

The first main consideration was the 'space between'. The starting point was the wide landscape, new to me, and I produced a simple pencil drawing of the shapes of fields as defined by the trees and hedges.

Thinking in terms of development and adaptation to different materials, I simplified further and reversed the tone, using black felt tip — dark spaces, white divisions — with the idea that this could be the basis for a print, possibly on fabric.

Jackie Goodman
(Teacher)
Hessle High School
Hessle

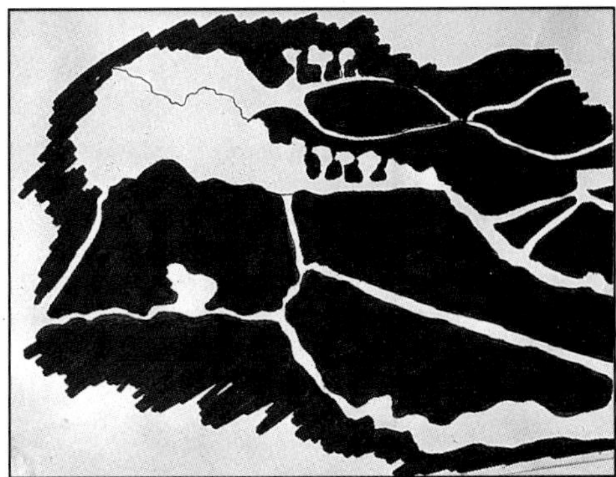

I then took the same section of landscape and added the man-made elements which I had previously omitted. This drawing was treated as a more narrative interpretation, using chalk and charcoal. The man-made elements — telegraph poles, fences — cut across the more natural landscape creating new and arbitrary areas of space. These could be treated in different ways; the composition could become a new arrangement of shapes, using the introduction of these elements or concentrating in detail on one of the identified constitutents of the landscape.

Although I began with the space that was the field or fields, other subjects were there to be used in the same way — the trees and their branches, the hedges. Pencil, pen and ink, water-based felt tips, water colour and pastel are the most suitable media for this approach.

Jackie Goodman,
Hessle High School, Hessle

Her description gives an excellent impression of how she tackled her brief. It illustrates how from one starting point, other approaches emerge, and how, while the drawing takes place, the thinking goes on:

The blue/grey smudge of distant hills, distinctly blue against the grey sky. I wonder what determined the shape and size of the fields.

Jean Hockenhull,
Riddings Comprehensive School, Scunthorpe

These examples prove once more how one hard day 'in the field' can produce so many ideas. It is back in the classroom that most of these ideas will be developed. The need for resources has already been stressed: postcards, pictures, artefacts, bits gathered from the day out and, of course, sketches and notes for reference. Displayed around the room, these can feed the stimulus of the original idea. The captured scenes can be re-created into some powerful forms, using images drawn at different times and perhaps exploring ideas which arose as thoughts rather than images.

Gavin Ward
Aged 13
The King Edmund
School
Rochford
Essex

Suggested activities

1 (a) Ask the children to collect a series of drawings showing how the edge of fields and spaces shows or determines the shape of the space (in other words, how they become lines across the landscape or environment).

(b) Within the field or space, there may be one object to which the eye is drawn, whether a tree, shrub, person, animal, machine or simply a change of colour in the vegetation. Ask children to collect a series of sketches showing such points. Bear in mind that an alternative is that the edges of the space or a group of fields may converge in such a way as to focus the attention.

(c) Ask children to note in their sketches the varied and different shapes created by fields and spaces.

(d) The convergence of the edges of these fields and spaces, and the folds and undulations within them, may convey something of the three-dimensional form of the environment. Ask children to go out and identify these within their drawings.

(e) Ask children to make sketches of the patterns created by the linking together of fields and spaces, or by the alternative cultivation of different plants and crops.

(f) Fields and spaces, such as gardens, parks, and wasteland, all have different surface textures of which a visual record can be made.

(g) Similarly, different tonal qualities can be observed, reflecting the vegetation, surface, fall of light and the variation in surrounds. Sketches, perhaps using carefully organised worksheets, should be made to show these.

(h) Observation of colour can be used to extend many of the above suggestions.

2 Make a model, on a largish scale, of a group of fields or an area of parkland based on an identified site and use the model for a series of further drawings, paintings and imaginative work.

3 Identify an accessible space and make a visual record of all the significant elements within that space. This could take in seasonal changes. (The work produced could form the basis for activity 2.)

Trees, Shrubs, Flowers and Grass

This area of interest is likely to be the easiest to explore within the context of the classroom. Many schools have trees outside their windows, and leaves, weeds, grass and flowers are usually at hand for children to draw. Providing simple materials such as pen and ink, water colour or soft pencils are used, such drawings will appear to be largely successful. However, a common problem is that children often work on such a theme without making any real reference to the organic forms before them: trees are drawn as cabbages on the top of sticks; flowers reflect the children's memories of the caricatures in their picture books. The experience of drawing has been simply that of keeping the children busy and the pictures fail because the need to get them to 'look' has not been properly met.

> **He began to paint the sky. We talked about colours while he was doing the picture; the colours of the sky, the land in winter, and the trees. Frank often gets confused over the names of colours but he knows which ones he wanted to use. The tree came last and Frank went outside to look at a tree before painting it.**
>
> Karen Hopkins,
> Whittlesea School, Harrow

The dialogue between the teacher and the child can be most significant. The teacher begins to create the scene, giving clues and hints as to what the child should be looking at, all the while enthusing him or her toward the subject.

> **Before painting this picture we talked about winter and the things we see around us. We went outside, looked at the trees, the frost, the sky, and felt the cold.**

The importance of language

A function of education must be to help provide children with a language which enables them to recognise and understand the environment around them. They should be equipped with the knowledge and the means to describe the surrounding architecture, landscape and fauna. They should be able to articulate the 'fine detail of the small place'.

> **Taking into account the school environment, I decided to base the study on trees, involving first and second years. . . . The**

In this drawing, it is the pattern of the bark which has been recorded
Pupil aged 8
The Chiltern Primary School
Hull

students made lists of as many different types of trees they could think of. We compared notes and went outside. The school has an interesting assortment of deciduous trees: all that remains of a once flourishing orchard plus Horse Chestnut, Sweet Chestnut, Willow, Birch, Ash, and others. That was our starting point. We looked at the variety of shapes each tree formed, the different barks, the variety of leaf shapes, and collected, pressed and drew as many different kinds as we could find over the following weeks. . . . Later on in the term we looked at trees painted by famous artists — particularly Van Gogh, Constable, Pissaro and Corot. Some of the children chose to write some poetry on trees, others preferred to paint owls, foxes, caterpillars and other woodland creatures. . . . Finally, the groups did some close-up drawings, enlarging sections of bark and making collages from their studies, using grit, string and wool.

Heather Sturdy,
The Commonweal School, Swindon

This teacher's approach illustrates how looking at trees and shrubs can direct attention toward specific detail, while at the same time open out a drawing project into the wider curriculum. There is an immediate connection with natural science, while the problem of how to approach history of art is also partially resolved. A good resources/visual stimulus collection should comprise not only pictures and pieces of trees, bark, twigs, leaves, etc., but also examples of the way other artists

A carefully organised painting can be produced using a combination of sketches made in the field and in the classroom. In this painting, the badger was drawn from a model

Dean Berry
Aged 16
Minsthorpe High School
Pontefract

Alison Willi

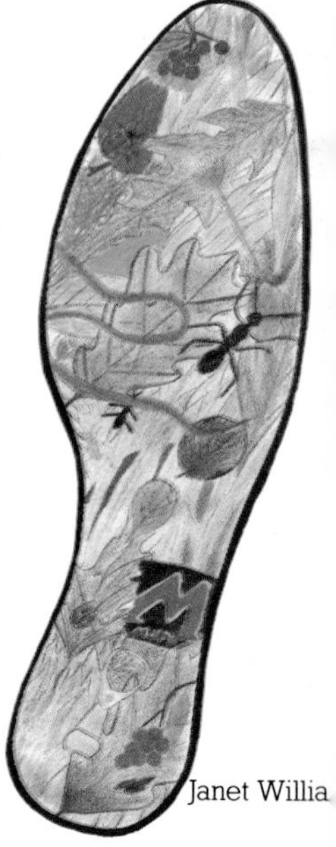

Janet Willia

have seen the same subject matter. Children learning the visual arts too rarely study the work of other artists, as they do when exploring the arts of poetry, music or prose

Critical studies

While researching the Schools Council 'Arts and the Adolescence Project' in the mid-Seventies, Malcolm Ross noted the frequent failure of the secondary art teacher to use the work of previous artists as reference points or examples. It is no different in the primary school, where children may often be asked to describe what they can see in a commercial photo-

graph or example of graphic art but are all too rarely asked to do the same with an established work of art.

Youngsters need to have their attention drawn regularly to existing examples of the visual arts in order to grasp something of the social and cultural climate which has shaped certain artists' particular vision. The various areas of interest explored in this book can be used as a starting point to encourage children to look at artists' work in this way. The skies of John Constable, the fields of Van Gogh, the machinery of Wyndham Lewis and the figures of Stanley Spencer are some obvious examples.

Asking children to note the way an artist has chosen to say something about his/her environment and what is going on within it can be a fruitful exercise. Refer to specific artists, as indicated above, or invite the children themselves to identify examples of work by other artists. It might be worth encouraging pupils to include postcards, small prints or photographs when they are asked to gather information on a theme. The areas of interest provided in this book could be used as a simple way of searching out the work of established artists or as a means of identifying work on a gallery visit. Some reference may also be made to established works in Part II, where concrete examples should help pupils grasp the abstract concepts involved.

Christine Smallman

Sara Owen

Here children have imaginatively re-created pictures of the vegetation that might be crushed by their footprints in a field

Pupils aged 11–14
Lewis Girls Comprehensive School
Hengoed
Mid-Glamorgan

'Entrance to a Lane' by Graham Sutherland

Classroom work

Strategies for working in the classroom worth considering include the use of twigs and branches as models for 'trees'. Even small 'landscapes' can be created using a range of foliage. (It has been said that the trees of Thomas Gainsborough resemble broccoli stalks because he frequently used small model landscapes with broccoli stalks, lit by candle-light, as his subject matter.) Similarly, animals and insects found in foliage can be sketched from the range of stuffed and preserved creatures available from museum loan services.

Flowers, weeds, foliage and grasses are the easiest objects to obtain and place in front of children. As the child studies and draws the line and form of a small plant or flower-head, so he becomes aware of natural development and growth. Thirty children drawing 30 leaves and picked flower-heads produce a massive visual aid for future work in science, as well as developing their drawing skills. The heads of large cut flowers or discarded vegetable tops provide similar scope.

The quality of the object to be drawn is important. It should be complex, both in form and colour. Contrast of shape and colour will intrigue and stimulate the pupil's interest and lead to an intellectual involvement with the object. Children become easily bored with simple shapes, perhaps because any comparison between the drawing and the simple shape too easily reveals inaccuracies, whereas a complex subject is too difficult for anything but cursory similarities to be noted.

Elaine Young
Aged 14

Catherine Jackson
Aged 14
Henry Cavendish
School
Derby

Suggested activities

1 (a) Ask children to note the linear qualities to be found on bark, leaves, branches and flowers, and amongst different grasses tangled together.
(b) Make sketches showing how one tree, shrub, flower or clump of grass can be identified as a focal point within a particular view.
(c) Ask children to record the different shapes of trees, shrubs, flowers or grasses.
(d) Certain leaves, patches of grass, shrubberies or woodlands will convey a sense of three-dimensional form. Children can be asked to identify these in the sketches.
(e) Close observation of the foliage of plants, shrubs or bark can reveal an exciting range of rhythmic patterns. Build up a collection of drawings showing these. A similar exercise can be done noting texture rather than pattern.
(f) Varied tonal qualities can also be seen and noted on different forms of vegetation or grasses within fields and other spaces, woodlands and shrubberies.
(g) Variations and contrast of colour can be observed by extending activity (f) but especially in working from flowers. Children can be asked to collect visual notes on families of colour, and to identify different colours or contrasts in colour.

2 Dig up a sod of earth and grass and use it as an object to draw in the classroom. The resulting drawings could act as a model for further paintings or for three-dimensional forms, perhaps with imaginative creation of a world beneath the surface.

3 Create some dense foliage in the classroom using a range of houseplants, tomato plants, runner beans etc., according to availability. Work on this could range from simple reproductions of the foliage to imaginative re-creations of jungles and primitive forests.

4 Use card, tissue and any available scrap to make (in groups or individually):
(a) a model garden
(b) a small wood
(c) a unique foliated area.

5 Using paper, corrugated card and the cardboard tubes from rolls of fabric and carpet, make a large-scale 'foliated' area in a corner of the classroom. Use it as the backcloth to other drawing, painting and modelling work, and to creative writing and imaginative excursions, and/or to house an exhibition of other environmental work.

The thin line of buildings controls th whole landscape in t picture

Janette Patterson
Aged 12
Ryecroft Middle Sch
Uttoxeter

Buildings in a Landscape

Barry Whiting
Wolfreton School
Willerby

The linear pattern of hedges in an undulating landscape is more evident and the farm buildings and cottages act as focal points or resting places for the eye when viewing a panoramic scene. Very often the ribbon hedges and lines of trees lead you to the focal point in a very natural way. . . . The temptation which has to be resisted is to over-emphasise the detail of the hedges and 'lose' the focal point. I like to use them purely as interesting direction lines to draw attention to the subject. . . . Against the background of fields, buildings tend to be seen as areas containing colour rather than stark shapes. Seen against the sky, however, the same buildings become silhouette shapes, which are a stark contrast to the softer shapes of the trees. . . . Moving into a townscape, individual buildings are more difficult to record as they are seen against a background of similar shapes. I found that the rooftops and chimneys were of more interest but it was a problem finding a suitable viewing point.

Barry Whiting,
Wolfreton School, Willerby

Careful observation c semi-detached house may reveal differenc rather than similaritie

Rachel Askwith
Lisle Marsden Middl School
Grimsby

In areas such as the flat plain of Holderness or the wide expanse of the Fens, individual or small groups of buildings take on a powerful significance. They act as a focal point and draw the attention, while the lines of their architecture gently re-direct the eye back into aspects of the landscape. The drama of such buildings can not only fire the enthusiasm of the child to capture the image on paper but it can feed their whole imaginative response to the area, which goes beyond the purely visual. In other words, creative writing is likely to be as much an outcome from this approach as drawings and paintings. The direction of such imaginative work depends upon the lead given by the teacher.

For me the day would seem to have reinforced my idea that it is not the quality of the media that is provided which determines the quality of the work produced but rather the wealth

of the experience — the breathing, seeing, enjoying, enthusing, sharing and feeling a part of the environment. This is how I would try and approach it with young children.

Karen Hopkins,
Whittlesea School, Harrow

But that 'breathing, seeing, enjoying, enthusing, sharing and feeling' can probably only occur if the individual has the confidence to allow such a reaction to take place within themselves — and that confidence comes from knowing what to do, feeling able to do it and wanting to be involved with it. Direction is all important.

Best then to look for something simple — buildings in isolation — and concentrate on the basic shape. . . . I need to feel, see and know a place alone before I can enthuse. Walking along the top road, I felt rather empty about the whole idea. The first drawings reflected this. It was a cold morning. . . . Once down in the bay, the enthusiasm and excitement began to grow as I began to feel the atmosphere of the place. We spent much time scrambling up and down steps finding new surprises and excitements and the coldness began to go. Barry was taking photographs. It became a matter of what not to photograph — there were so many beautiful shapes. A large bird of prey hovered over the bay, still against the gale. In that stillness, I found incredible strength. . . . One of the images I remember was seeing, through a haze of rose-bay willow-herbs in seed, a group of cottages huddled as if behind a feather curtain. It was magical. . . . After lunch, we were off in the car across the moors. This was simpler: a small farm here, a group of dwellings there. Definite shapes in a distant landscape, blurred by the wind and the weather; subdued colours but still a patchwork spreading as far as the eye could see.

Julie Walton,
Grange First School, Grimsby

Two examples of the way original sketche[s] colour studies and photographs have be[en] combined to re-crea[te] impressions of the vi[sit] to Footdee

Norman Grey
Aged 14
Summerhill Academ[y]
Aberdeen

Reference throughout this book is to the landscape but that word is intended to cover the whole external environment — townscape and seascape as well as 'land'scape.

The discovery of oil has revolutionised the city of Aberdeen in many ways. It has one of the fastest developing industrial and housing estates in Europe. Yet in this city of change one thing remains constant — the sea. . . . The pupils were asked to list items affecting their lives (e.g. school, work, parents, housing). Employment and unemployment seemed to figure largely and this led to a discussion of job availability in Aberdeen. This in turn led to a discussion of traditional and new fields of employment in the city. An area of the city that reflected traditional trades while being surrounded by modern development was the tiny fishing village of Footdee (pronounced 'Fittie'), tucked away in a hidden corner of the harbour. The harbour itself has been greatly modernised and evidence of

the booming oil industry abounds. Strangely, the little village has been beautifully maintained and its original character of hardy self-sufficiency preserved. The class opted for a study of the village and four lessons were reserved for the collection of data.

Week 1	**Discussion of topic.**
Week 2	**First visit to Footdee and first pencil sketches.**
Week 3	**Continuation of pencil sketches and first batch of photographs.**
Week 4	**Photography and 'texture' collection, pencil and water colour sketches.**
Week 5	**Pencil, coloured crayon and water colour sketches, final photographs.**
Week 6	**Review of material and selection of developmental areas.**
Weeks 7–12	**Assembly of material into finished products using the following techniques.**

(a) Collage (reprographic/photographic) emphasising textural pattern of natural building materials.
(b) (i) A symbol for Footdee (collage, paint)
(ii) A poster for Footdee
(c) A flag for Footdee
(d) Pen/ink/wash drawing/collage
(e) Wax resist/pen and ink
(f) Texture rubbings

Tina Stockman,
Summerhill Academy, Aberdeen

Alison Scott and
Billy Kemp
Aged 14
Summerhill Academy
Aberdeen

Once again, the response of this secondary school indicates the potential richness of experience and the range of activity that can be generated by such a visit, when armed with adequate facilities to record impressions. The unstated but crucial element is that of personal enthusiasm, and the way each weekly activity was supported by dialogue with the teacher and the organisation of the school.

The natural organic form of the trees is shown against the measured balance of the buildings

Aged 11–14
St. Mary's Music School
Edinburgh

David Horne

Donald MacBride

Contrasts

Town buildings framed by parks, gardens, churchyards and derelict sites provide an infinite source of ideas. The contrast between natural forms and architecture is an obvious starting point:

> **Stourbridge is a small town west of Birmingham. It is on the edge of the Black Country and the industrial Midlands but also borders onto the green belt countryside. . . . Stourbridge, like many other towns, has expanded to trespass on reserves which previously were unaffected by man but inhabitated by many different forms of wildlife.**

L. P. Dukes,
Redhill Comprehensive School,
Stourbridge

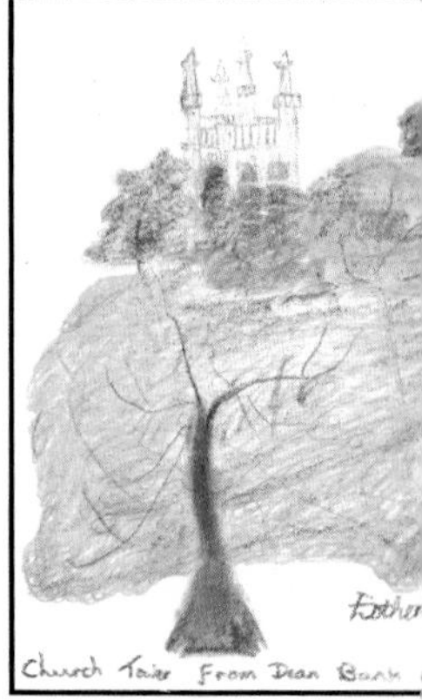
Esther Geldave

Suggested activities

1 (a) The line surrounding a group of buildings or a line within a structure itself can be noted, as can the way a long industrial building or terrace of houses may create the same sensation as a line.
(b) One building or a group of buildings may be a dominant focal point in the environment. Ask children to collect visual examples.
(c) Children can be asked to make a series of sketches tracing the line around various buildings and structures, and to note the shapes revealed. They can also note the shape of windows, doorways or other elements which go to make up a building.
(d) The shape of a particular structure or the way a group of buildings relates together may convey definite three-dimensional form. Ask children to find and make visual notes of such places.
(e) Ask the class to note the way a set of windows, doorways or chimneys, a collection of buildings or the repeated design of particular houses can all communicate a certain pattern.
(f) The different materials in one or a number of buildings, or contrasts between the texture of one structure and the surrounding environment, provide much scope for observational work.
(g) Differences between buildings or areas of the same building, and contrasts with the surrounding landscape as shown by changing weather conditions, the time of the day or year and changes of light, will all produce a range of tonal qualities which can be observed and recorded.
(h) Colour exercises can extend the above activities. Similarly, observing the contrast between natural landscape and painted buildings, and the differences in building materials, and between painted, weathered and bare buildings is another valuable exercise in visual notation.

2 Select a specific group of buildings, such as a small industrial estate, a collection of houses, a farm or a village, and make visual studies of all the significant elements to be seen within or around the area.

3 Make a model of a small building within a landscape, for example, a park shelter, a shepherd's hut or a seaside stand, using a pliable material such as clay.

4 Make a visual record of the changes that take place in a building and landscape throughout the seasons.

5 Possible themes for imaginative work are:
(a) The gaunt and lonely building.
(b) The terrace looked out across an overgrown unkempt garden.
(c) Against the skyline, the factory breathed a sense of activity.
(d) Standing there, the building changed the whole landscape around it.

Stone, Quarries, Pebbles and Cliffs

Stones and pebbles are usually readily available. Children can be invited to collect their own and bring them the next day. Small still life arrangements can be made. Patterns, shapes and strange formations can be put together and drawn. A permanent collection of stones, which also includes a range of interesting minerals and fossils, can be kept at school.

The quarry and the cliff face lend themselves to the school visit. The advantage is that, within a comparatively small area, there is a good deal to see, touch and draw – and the confines of a quarry or the bottom of a small cliff face can make a group of children easy to control and direct.

Initially the children were encouraged to collect and point out visually enticing subjects, such as plant forms (alive and dead), rocks, pebbles — indeed, anything they found visually attractive or stimulating — and to sketch or photograph views and scenes that captured a particular mood or emotion. . . . The children, where possible, were encouraged to handle the objects (so as to foster an appreciation of texture, weight and outline) and, where appropriate, to observe objects and views from all angles, by walking round them, or from unusual angles, or to study a small section in close-up. . . . At the abandoned slate quarries the children were encouraged to look at the range of colour, and the texture and pattern of the

Roger Henry (Teacher)
Brumby
Comprehensive School
Scunthorpe

Beverley Mawer
Aged 11

Mersey Junior High School
Hull

sheer slate quarry sides, at the shape and lines of the train track, chains, pulleys and disused rusting machinery, and at the arrangement of roof tops, chimneys and work sheds crouched at the bottom in huddled groups, as well as to feel and absorb the grey bleak loneliness, harshness and desolation, the overpowering and over-awing scenery and its dominating effect on the people who struggled to survive there. . . . Later we looked at our finds and records, while our experiences and impressions were still vivid and fresh, and discussed why they were so attractive; for example, the grey wetness of towering slate waste pipes against bleak, vast, sweeping skies, the intricate curves and swirls found in tree barks, the lush variety of the many shades and hues of the greens and rusts in the mosses and lichens, and the clusters, patterns and arrangements they formed against the greyness and solidness of the stone wall itself. . . . The children were actively encouraged to discuss what they were trying to achieve and any problems they were encountering; to discuss different materials, media and techniques to achieve specific results; and to ask for guidance where needed. During this work vocabulary was developed, and alternative techniques were introduced as an integrated part of the practical work. . . . At all times the children were strongly encouraged to look and carefully observe and to become imaginatively involved — with the experience of holding a moss-covered stone, of stroking bark designs, of feeling the loneliness and bleakness as well as the magnificence and grandeur of the overpowering landscape.

J. Croney,
Westwood Park Primary School,
Eccles

Michael Rauley
Aged 11

Mersey Junior High School
Hull

The opportunity to touch and feel the objects is essential to the process of capturing their quality in the mind and on paper. And being in the landscape itself is as much a three-dimensional experience as touching the stone, the bark, the flower or the wall.

Imaginative work

Earlier, reference was briefly made to imaginative work. An example of this was that encouraged by a curriculum study group in Grimsby, who involved a number of their infant and junior pupils in activities similar to the above. They then asked the children to look hard at the rocks, stones and pebbles. Inside these forms they were asked to find caves and caverns. The original and exciting work which resulted involved elves and fairies, monsters and demons, and Tolkienesque stories.

The patterns of the stone structure are basically linear. Within this structure can also be observed the colours of the stones enriched by moss and litchens, which also serve to give surface texture. Look also at how the stones have been quarried and cut. Look too at weathering processes of erosion.

As a small item within a total landscape framework, a wall structure may seem relatively insignificant, but many styles and arrangements can be observed and translated in many media —

1) Direct drawing – line/
2) " " + painting – line + colour
3) Paint and collage – line, colour and surface texture
4 Fabric applique – as above but with decorative techiques of embroidery etc. to bring in the qualities of moss + lichen.
5) Ceramic construction techiques in making "wall pots".
6) Printing through cutting or small block

As additional resource, consider piles of bricks in their patterns when systematically piled on a building site, or when the pile is disturbed and accidental arrangements result. Examine carefully the language, both technical and biological.

Suggested activities

1 (a) Investigate, through drawings, the types of line to be found in stones and pebbles in quarries or cliff faces.
(b) Build on (a) to find focal points created either by surface markings and cracks or by contrasting elements in the stones or pebbles.
(c) Explore the infinite range of shapes to be found in the edges of pebbles and stones, as well as in the walls of quarries and cliffs.
(d) Note how the lines, points and shapes observed will sometimes imply the three-dimensional form of the whole.
(e) Different minerals and stones will have variations in texture which can be noted and re-created in drawings and sketches. Patterns can be revealed in the same way.
(f) There is scope for tonal studies in the play of light on the many planes within a quarry or cliff, or in the way different pebbles and stones reflect the light.
(g) The subtlety of colour in various pebbles and stones, the various earth tones of a quarry, contrasts in a cliff face and weathered stains running across stone are all subjects for colour exercises.
(h) Different stones, pebbles, cliff faces and quarries can have within them a wide variety of textures.

2 Take a number of pebbles and/or stones and arrange them in a varied 'still life', i.e. placing some close together, some separately and some in piles. Concentrate on the spaces between the objects and seek ways in which to show these spaces in drawings and paintings.

3 Reproduce the shape, form and texture of stones and pebbles in clay.

4 Quarries and cliff faces often have a powerful effect on the visitor. A valuable objective and subjective experience would be for children to make large-scale direct paintings of them on the spot.

5 Possible themes for further imaginative work are:
(a) Between the stones and into the caves.
(b) Grim and unyielding, the rock face stretched above us.
(c) The men stood quietly waiting in the quarry.
(d) Pebbles glistened in the water beneath the cliff.

s work shows how,
:e again, the process
:ts with looking at
l recording a small
ect of a specific
me and grows from
starting point

n Thirlwall
viser for Art and
aft
derland

The Sky

In any study of the landscape the fate of the sky closely parallels that of the postman in one of G. K. Chesterton's famous Father Brown stories. The postman is the figure in the street whom nobody noticed because 'postmen are always there'. In fact, of course, he turns out to be the key figure, the murderer. He is the background figure who is largely responsible for the drama. Similarly, the sky is rarely the focus of people's attention, yet it provides the dramatic backdrop which sets the scene and highlights the other features.

Asking children to observe the sky is likely to prove an unsatisfactory experience. It is largely seen as being in one of three states: blue, grey or black – with the addition of 'cloudy' as a temporary condition. Yet anyone looking at the sky for some time will find something rather different. Cloud formations are frequently substantial, with recognisable form. Colour and tone provide gentle frameworks. There is a sense of permanent rhythm which fascinates and soothes. The difficulty lies in believing that something of these qualities can be captured on paper.

Just as it became clear that fields and wide open spaces can draw attention to the objects that surround them, or that these objects can give meaning to the space itself, so it is with the sky. And similar problems arise. How do you define which is the sky and which the objects that fringe it? Clouds cover hills and hills rise against the sky. Tall buildings and objects reach up into the sky while it presses down hard against and around them.

At one stage, the process of 'skying' was an essential part of the education of an artist. They were encouraged to explore its depth, colour, shape and movement and make visual notes. It is both desirable and feasible that children should be invited to do the same today.

Weather conditions:
a.m. Cool, overcast. Limited tonal content in sky.
p.m. Windy, mist, rain. No tonal contrast in sky.

Observation Area:
a.m. Robin Hood's Bay. From the beach, looking toward Ravenscar.
p.m. The moors between Whitby and Fylingdales.

A sky with little tonal contrast and no recognisable formation can still present a worthwhile but difficult study

Harold Gosney (Lecturer) Grimsby College of Technology

General Aims:

To relate the sky to the land masses. Insufficient tonal contrast in the sky alone. Difficulty experienced in starting drawing too early — in order to get a good perception on landscape/sky relationship needed to take in atmosphere for longer period of time. Perhaps walking and looking for a day — then produce drawings.

a.m. Ravenscar. The obvious simple shape (especially on first visit to the area).

Used boxes to:

(i) Consider balance between sky and land mass

(ii) Experiment with drawing media.

Photographs — to hopefully back up studies.

Essential to simplify shapes, i.e. land mass was basically horizontal and Ravenscar, although impressive, was a relatively small vertical form. The sky was overwhelming — much more than just a backcloth.

Black conte crayon severe on white paper.

White chalk might have helped to produce better tonal statement.

Sepia conte pencil seemed to relate better to white background.

p.m. Conscious of strong sculptural shapes on the moors. Sky still has considerable impact but weight of the land mass seemed more important. Tonally the sky closer to the land mass than during the morning. Some cloud moving fast though not having big visual effect on perception of land. Distance appearing to change as mist came and went.

n.b. Landscape starts from point of vision. Mistake always to set subject matter in distance and to ignore foreground.

Tonally — nothing in land mass as light as sky, even on a very grey day. Unable to produce subtlety of tones between fields, and hedges and so forth in the time available.

Harold Gosney,
Grimsby College of Technology

Harold Gosney, who writes of his difficulty in drawing the sky's tonal qualities, gives a revealing picture of his method. It is the method used by most of us, except that many pupils and teachers have neither the knowledge nor the experience of media and process to give such a detailed explanation. And, as stressed earlier, the more a person can articulate such a process, the deeper their understanding of it and the greater the possibilities of repeating and refining it.

The painting of th
comes from carefu
observation of col
and tone, but the
apparently unobs
remains the famili
caricature of a gre
patch

Janette Williams
Aged 5
Nunsthorpe First S
Grimsby

The changing cloud formations appear to alter the environment beneath them

Michael Adeney
Aged 18
South Hunsley School
North Ferriby

The change in the colour of the sky changes the light in the landscape

Alison Adeney
Aged 16
South Hunsley School
North Ferriby

Sandra Langford
Aged 8
Williams Middle School
Grimsby

It is the sky which draws particular attention to the buildings and the environmental furniture in this picture.

Suggested activities

1 (a) The often gentle and delicate lines which divide up the cloud patterns themselves can provide an intriguing subject for a series of sketches.
(b) Explore how a particular cloud or a 'foreign' element such as a plane, bird or insect, or even the way clouds converge, can give a special focal point to the sky.
(c) Ask children to observe and draw the infinite variety of intriguing shapes into which clouds both build up and break down.
(d) A three-dimensional effect is highlighted as clouds sweep and climb, combine and separate. Similarly their colour, tone and outline presents a sensation of form. Ask children to observe and draw this element of form in the sky.
(e) The sky has a multitude of subtle textures within it which children can be encouraged to note and reproduce in their sketches.
(f) The traditional 'grey' sky generally contains a mass of tonal quality. Encourage the patient and careful consideration which allows this to be appreciated and reproduced.
(g) Interesting exercises and sketches will result from concentration on the subtle tonal changes in the sky, as well as on the extreme contrasts between clear skies and clouded areas, between dark and light clouds, and between the sky and any selected foreground subject matter.

2 Explore ways in which a three-dimensional interpretation of cloud formations can be presented. Balloons, papier-mache and wire may be suitable for a large-scale sculpture. Clay, card and boxes will work on a more small-scale, individual level.

3 Ask children to work on a large scale and in the open, preferably painting wide expanses of sky and cloud moving across a gently undulating land or townscape.

4 Small-scale and photographic work can be done on the sky reflected in windows and on highly polished areas.

5 Possible themes for further imaginative work:
(a) Dusk.
(b) An approaching storm threatens the whole environment.
(c) Early morning light awakens the area.
(d) Through the window, the sky looks full of hope.

The Seasons

The seasons are an attractive area of interest, full of associations and potential for development. However, as a theme this is less easy to make immediate use of in the classroom. The title itself implies a contrast; it implies that the work will reflect changes between the seasons, which in itself means a process spanning a period of months. Also, the theme can appear so broad – almost abstract – that it may seem as potentially unrewarding as thrusting someone into a rich landscape and saying "Okay, draw!". Finally, it implies organisational and inspirational problems: organisational because work will need to be stored; inspirational because completed work will have to be re-started and similar topics repainted and redrawn.

However, the value of the theme lies in this very area of difficulty. It is important to learn to start again, to sustain interest in one idea, and to be involved in a process of reworking something. And looking for changes between the seasons positively encourages the observation of that essential fine detail.

A useful strategy is to highlight one area which is easy to observe regularly at particular times of the day and year. It may be a 'natural' area, in which plant life can be observed experiencing the changes of time, weather, temperature and light that comprise the seasons, or it may be the corner of a yard or building where light, shadow, colour and form appear to undergo a metamorphosis.

Recollection in tranquillity

This area of interest appears to lend itself more to the outside visit and the 'working from nature' approach. Nevertheless, classroom work can be important. Observation of the changes in the landscape can bring about an intellectual curiosity. Using the notes brought back, the observations, feelings and ideas experienced should be reconstructed as part of the school work. With the aid of the various visual, tactile and written resources collected together, the pupils should become involved in the process of re-creating original emotions by 'recollecting in tranquillity' what they saw and felt. Through seeking to create colours once seen, shapes once touched, sensations once felt and places once visited, the original experience is better understood. And not only may the places of the original evocation be revisited, but classroom work on that visit can also be remodelled, reprinted, redrawn or repainted to show changes in the seasons.

A work of art succeeds if it allows the artist to recreate the feelings which he or she originally experienced and which provoked the process of creation. Perhaps what makes a great work of art is if it allows others to come close to experiencing the same emotions as the artist.

Similarly, for children the work has succeeded if, during its execution and on completion, they sense something of their original experience and, by working through such feelings, gain in their understanding of them.

Again, the use of drawings and colour studies done on the spot can be successfully combined with classroom drawings to produce a considered re-creation of a particular moment in the environment

Jennifer Y. Holmes
Aged 18
All Hallows School
Bungay

A painting may illustrate the way the child feels about that subject. This big, heavy lumbering tree is as much an illustration of how the artist perceives it as it is a record of her skill

Lindsey A. Brown
Aged 17
All Hallows School
Bungay

Sixth-form pupil
South Hunsley School
North Ferriby

These pictures are attempts to imaginatively re-create the feel of a particular season.

Pupils aged 8–12
Willows Middle School
Grimsby

Suggested activities

Although some ideas are given here, in general the seasons will affect other areas of interest rather than produce their own series of exercises.

1 (a) A white winter may make it possible to explore line in terms of the black edge of fields or parkland against snow or frost. Looking at snow-covered trees will prove a useful line activity.
(b) Winter snow will reveal a footmark or a solitary stone as a focal point; in summer the single bloom in a flower border, or one field of yellow in five fields of green similarly becomes a focal point.
(c) A series of drawings could show how the shape of an organic form, such as a tree or flower border, changes over the seasons.
(d) A similar exercise would be to visually demonstrate how an area is changed, disguised, covered, stripped and exposed during the seasons.
(e) Observing texture provides similar scope: for example, the bulb which turns into the stem, into the flower, and finally into the skeleton.
(f) Identify some area or object of study which has distinctive patterns and observe how they change and evolve over the seasons.
(g) Find an area or object which contains sufficient visual material to study the changes in tonal and colour qualities throughout the seasons.

2 Make a series of large-scale paintings, prints or drawings showing how some striking feature in the environment undergoes major changes in colour and form throughout the seasons.

3 Build a three-dimensional model of, for example, a park, high street, seaside front or town garden and show how it changes with the seasons, resulting in four versions of the same model.

4 One large sculptural shape could be constructed such as a tree or flower border, which reflects the four seasons in different parts of the same model, i.e. different branches or corners.

5 Possible themes for further imaginative work:
(a) The landscape of a dying summer.
(b) Snow in the streets.
(c) Fires in a skyscape.
(d) Summer flourishes in a corner of the garden.

Machinery and Furniture

Often it is the machinery and furniture found in the landscape which gives a sense of meaning and purpose to the environment. The brightly painted tractor and the hay bailer draw attention to and indicate the use of the land. The train, lorry, narrowboat or car also focus the attention, providing action and interest, as do towers, posts and cables, signs and notices, walls, fences, gateways and so on.

This area of interest both provides something definite to look at, and places the environment into a specific context. Despite this, it is not a popular theme. One major difficulty is the amount of information a collection of machinery, equipment or furniture presents the artist with. At the same time, such objects tend to have a very particular form and shape, and therefore demand a high degree of artistic certainty.

However, one solution is to employ the strategies already outlined and to direct concentration on one element rather than the whole object. Part of a machine may be looked at for line or colour for example. Similarly, consider the texture and pattern in environmental furniture. It helps to have the objects in the classroom or to arrange a session amongst the objects.

Museums are well worth exploring:

The museum consists of the traditional dockside buildings, lock basins and canal barges, all in varying stages of repair.

Tina Sparke
Pensby Secondary School for Girls
Heswall

On the canal banks wild flowers and weeds festoon decaying machinery. The bright colours of the restored canal barges, depicting life at the turn of the century, contrast vividly with the rotting wood of the skeleton of the canal boats in the dock basins. The iron gantries, lock machinery and ropes all provide stimulating visual experiences. . . . Twenty seven girls from the fourth year spent an afternoon at the museum taking photographs, drawing, touching and looking. Armed with their information they set out to make a visual commentary of their experiences. Some chose to record, in abstract and visual terms, a colourful commentary of the barge restoration. Others, in inks and water colours, attempted to relate the restoration being carried out in the dock basins.

Gill Curry,
Pensby Secondary School for Girls,
Heswall

Work produced this way, perhaps using prepared drawing sheets, can be taken back to the classroom and used as a basis for further development: farm machinery can be placed in fields, industrial equipment into factories and barges into canals, for example. These kinds of objects provide the contrasts in colour, line, form, texture, rhythm and pattern that provoke visual interest. The vivid colours of a machine against a tranquil landscape; the harsh horizontal and vertical lines of equipment against the gentle planes of road and track; the starkness of a notice against the softly merging tones and earth colours of a weathered wall – these upset the traditionally conceived image. The very contrast can shake us into seeing with a fresh eye, and encourage the highly personal response.

Industrial archaeology does not offer such extreme contrasts but it does provide an opportunity for varying perceptions. Abandoned machinery and rusting equipment provides material for tonal and colour exercises, as well as stimulating thought and work on reconsideration and recreation of the past.

The natural colours of the wood, the rotting iron machinery, the areas of brickwork, their surface regularly broken by lines of mortar and grass, offered textural contrasts for observation, analysis and translation into drawing. . . . The shapes and forms of lock gates, walls and rotting piles were often immersed in dark shadows. The shadows concealed forms and the girls were encouraged to exploit the spacial ambiguities rather than the clear exposition of precise reference points. Reflections and distortions of the water helped some to recreate a sense of drama in their drawings. . . . The museum offered some complex subjects involving geometric and freely shaped objects in a situation where atmospheric tonal changes were observed. . . . The girls translated, directly from observation, their visual experience of the reclamation of a large decaying area of the environment.

Pupil aged 7
South Parade First
School
Grimsby

The visit to a local museum provides the opportunity to look at a 'machine' which was once to be seen in local streets. Back in the classroom, the combination of such visits to both the museum and the street can be used to imaginatively re-create scenes from the past.

Richard Gilbert
Aged 8
St Mary's First &
Middle School
Grimsby

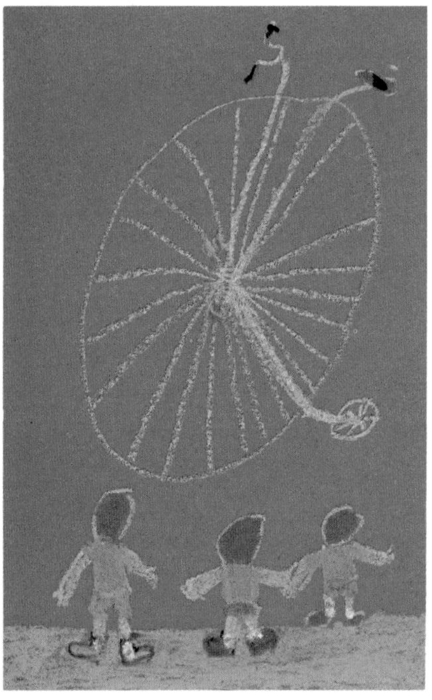

ENGLISH LANDSCAP
PROJECT — MACHINE

Looking
the area
somewhat
at machi
natural b
time hu
it became
in which
being very
provided s

spinner left in a field reminded me of similar stark sh
upwards into the sky. A possible starting point for
sculptural forms – print making etc.

The objects are a good source of inspiration themselves. the
between. Also:- positive and negative images – complimentary

Looking through a series of shapes – a good basis for tonal com
(pop-up). card construction) or ceramics
the situation in a different med
the whole and using in a

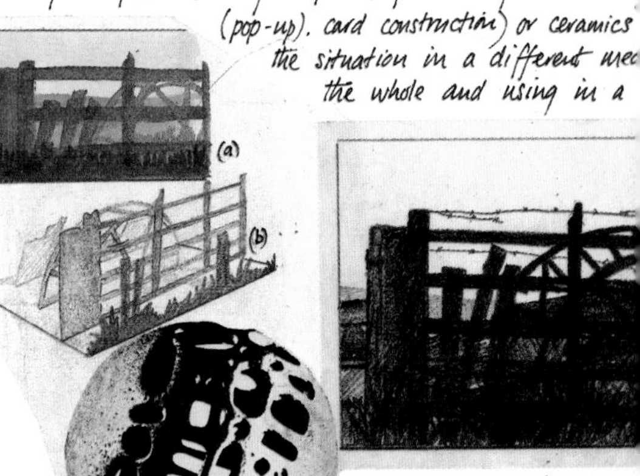

Ceramic sphere using the c
shapes to others beyond, as
new patterns, just as the
location did.

Maybe use different arrangement of shapes - textures - colours work well as glazes

Working parts of machines:– Interesting arrangement of shapes - colour. How did the machine work?* Could use as starting point for design work - ceramic panel - good colours here - ageing has toned down the brash colours to blend in well with natural surroundings. Left on their own they become overgrown and eventually are taken over by the landscape - a reversal of roles, machines used to shape and form the landscape. 'A battle!'

*Machines with unusual characteristics - Heath Robinson. fun machines - Emet. Surrealist constructions Junk machines.

'scape in
Bay it seemed
time looking
full of
nding much
s locations
see the ways
es, although
their nature
old potato
utting

3D work (b)
trying to create
tracting part of

through
d, it creates
the above

Repeated shapes and forms - possibilities for development as repeat design fabric print. Kaleidoscope or mirror the shapes and adapt images - different colour combinations.

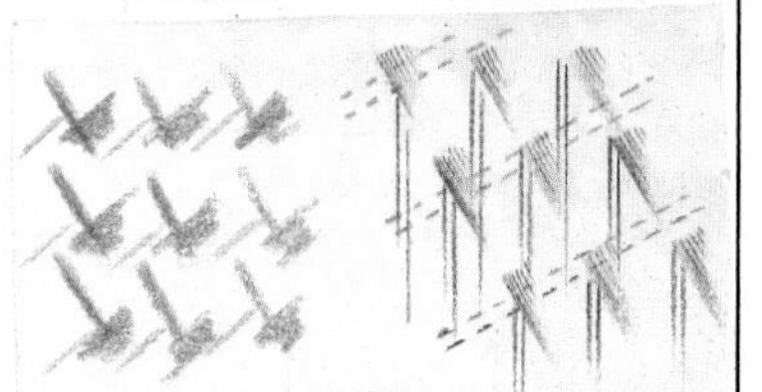

Summary - There are many possibilities to be tried and developed from the source material and as such may be it would be an idea to use it as a separate project. If it should be included as part of the whole, then the situation becomes the important feature ie. How the machinery blends or stands out in the landscape and the part it plays in it. As a purely personal reflection on this, what impressed me the most was the way the machinery was overgrown and abandoned, the feelings about the atmosphere inside the barn or farmyard - or the activity going on when machines in use - rooks and gulls working after the plough. The changes in the landscape between cut and uncut corn or ploughed land in furrows. I probably concentrated on one type of machinery (farm) - there are others - transport in general - building and demolition etc. Maybe I am just a romantic! Roger Henry

The teacher has made his own worksheet, carefully combining thoughts, observations and sketches in such a way as to help re-create his experience of the day in the landscape

Roger Henry (Teacher)
Brumby
Comprehensive School
Scunthorpe

Suggested activities

1 (a) There is plenty of potential to observe lines. Machines, engines and environmental furniture are often full of lines which take the eye in a variety of directions, which divide space or even describe the use.

(b) There may be focal points on the machine or objects, created by the apparent movement of lines, special markings or colour. Equally, the machine or environmental furniture itself may stand out as the focal point within an environment. Ask children to collect drawings of both kinds of focal point.

(c) The shape of such objects within the environment can be striking, like a piece of isolated sculpture. Ask children to identify and record such shapes, paying special attention to the way they affect the space around them.

(d) Many of the objects recorded above will have elements about and within them that convey a definite form, i.e. three-dimensional shape. Ask the children to record these.

(e) Further themes for drawing are to pick out and convey the rhythm of a group of machines, repeated shapes or marks within an object, or the balance within a machine or piece of environmental furniture.

(f) The broad nature of this area of interest means that it contains objects with an almost infinite variety of surface textures for children to record.

(g) Similarly, the range of materials and the way they absorb, reflect or diffuse the light will produce a wide range of tonal qualities to be observed.

(h) Colour exercises will be appropriate in any of the above activities. In particular, (f) and (g) lend themselves to valuable exercises in understanding the quality of colour. A further suggestion is to observe the effect of colour objects placed in an apparently alien environment, bearing in mind that this is often how machinery and environmental furniture is perceived.

2 Using cardboard and gummed paper, reconstruct a piece of machinery or environmental furniture. Use the model for further drawings and associated imaginative work.

3 Objects in the landscape often become an almost indistinguishable part of the environment, for example, an abandoned rusty tractor or derelict car. Find such an example and make sketches on site for work to be completed later in the classroom.

4 Small objects such as tins, iron implements and other bric-a-brac found in the environment make useful objects to draw and paint.

5 Possible themes for further imaginative work:

(a) The machine stood in the field looking bright and violent.

(b) In the grounds, a forgotten engine rusted into the landscape.

(c) The boat stood stark on the beach.

(d) The sign at the corner stood reaching into all directions.

Animals and Insects

Birds in flight or on branches, dogs in streets or curled up in corners, farm animals grazing or tied up in sheds, familiar mammals half concealed in their habitats, all these give the environment a sense of meaning and purpose as well as excitement and interest.

We decided to focus on individual animals or birds and their surroundings, making particular studies of colour and texture. As a result we embarked upon the exciting project of experimentation with all the drawing and colouring materials to hand. . . . We revived an interest in leaf and bark-rubbings, using wax crayons on a variety of papers, which in turn encouraged a more mature consideration of the creatures' supportive environment. The children discovered that over-rubbing textures effectively captured backgrounds of plant growth. . . . The following progression was to over-lay wax colours, and, in scratching off the surfaces, we discovered a suitable medium for ground cover. . . . A discussion about body covering led to a choice of 'soft' or 'strong' quality of colouring materials and many challenging experiments to achieve the 'feel' of fur, hair, feathers. . . . We established contact with the Schools' Museum service in order to borrow stuffed specimens of some elusive creatures for closer studies.

E. Drew,
Oratory Prep School, Reading

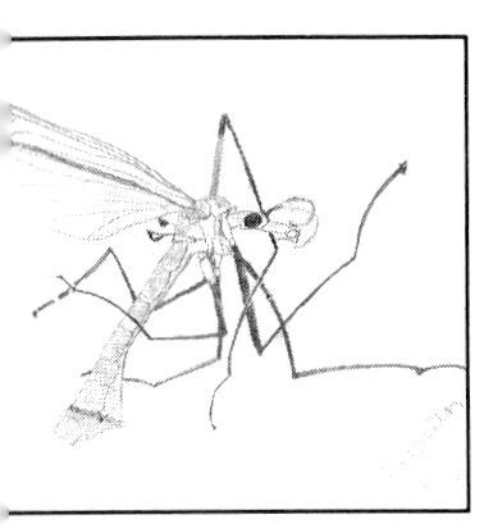

Drawings can be made from available exhibits in the classroom, both as a means of looking hard at the insect and to store information for further visual developments.

Pupils 14–15
Gillott's School
Henley-on-Thames

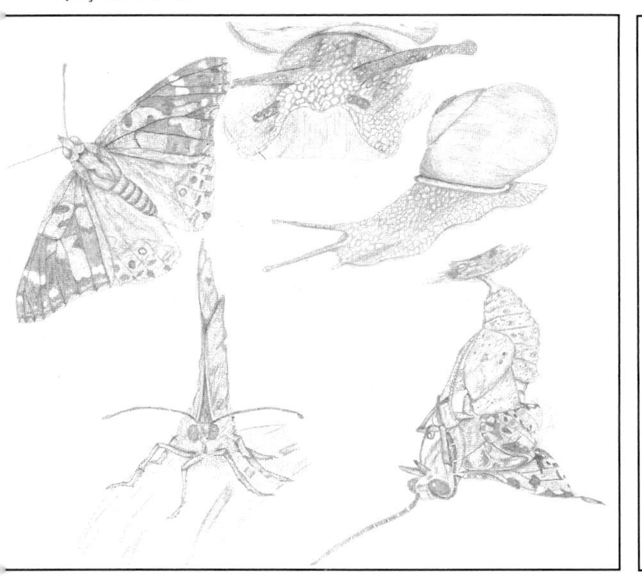

A practical problem is, of course, that most animals and insects rarely stay conveniently still. One solution is to make use of stuffed and preserved creatures, found in most museums and galleries. The local museum and education authority may have loan collections which include various wildlife. Some schools are able to build up their own collections.

The value in working from such exhibits is that they are three-dimensional and provide an infinite number of viewpoints, whereas an illustration or photograph provides only that of the original artist.

Drawings and sketches collected on visits to museums or galleries can be taken back to school and used in combination with other work to place the creature in a context, which could be its natural habitat or some dramatic or intriguing scenario.

The boys and girls involved are between 12 and 13 years of age, and we started the project by discussing the wild life most commonly found in the village and the surrounding area, coming to the conclusion that the seagull population outnumbered all other forms of wild life.... We decided to become detectives, following this boisterous form of wild life to its various haunts in order to discover more about its habits. This indeed we did, and the children were surprised by the aggressive, raucous personalities of gulls as they followed their busy routine.... Our project took the form of visits to the beach, lighthouse and surrounding rocks for both sketching and photographic sessions. We also noted the visits the gulls made to the fun fair, with its famous corkscrew ride, now deserted during the cold north-eastern winter, as well as the market stalls, after the market has closed, and the fish quay in nearby North Shields.... Closer to Shiremoor, on the farms, we observed the seagulls following the tractor as it turned over tasty morsels for their lunch. Closer still, the gulls are frequent visitors to our school fields, where they sit motionless as though inspired by Alfred Hitchcock, until the tide turns, or conditions improve sufficiently for them to resume their fishing trips.... Once all the information was gleaned from their observations and sketches the pictures were started. The children chose the media they felt most suitable for their subject. They used their photographs to refresh memories and provide details, resorting to reference books when details of the birds' anatomy caused problems.

In my opinion the project has heightened the children's awareness, not only of the gulls but the places they visit, their environment, in particular the rock formations, which has led to many creative ventures within the school curriculum.

Mary Austin,
Shiremoor Middle School,
Newcastle-upon-Tyne

Clearly there are exceptions to this way of working. Some animals move slowly or remain still for a reasonable time. Farm

These pictures illustrate how a theme may be undertaken in terms of finished work. Using the combination of visits and classroom work, the different environments which the seagull inhabit have been explored. Resources such as sketches, photographs and illustrations have been used as reference points. At the same time, discussion about the habits and life of the seagull constantly fed the activity of painting.

Paul Gouldsbrough
Aged 13
Dawn Thompson
Aged 12
Neil Gow
Aged 12
Lisa Hardy
Aged 12
Shiremoor Middle School
Newcastle-upon-Tyne

animals are obvious examples, as are tortoises, fish in tanks, and classroom pets such as guinea pigs, rats and rabbits. Chickens, duck and geese have also proved successful and exciting visitors to the classroom — and there have been stranger examples!

> **During the week all the pupils were taken by the leader into the limestone caves. They were kitted out in waterproofs, helmets with lamps and led underground for a considerable distance, sometimes up to their waists in water, crawling through tunnels and along ledges. . . . At the end of the week all of them wrote descriptive passages of the caves and of their feelings and emotions. . . . To any form of life, caves represent a most unusual habitat and in fact provide one of the most distinctive of natural environments. . . . The children had been asked to switch off their lights when underground and they stood for a while in complete darkness and complete silence — it was not difficult for them to understand that it is the darkness which is probably the most far-reaching and influential of all the peculiar characteristics of the cave environment. . . . The only animal the pupils were actually aware of in the caves were the bats, especially one pupil who was frightened by a bat he had disturbed flying straight into his face! As there was so much interest aroused in caves I've begun a special project building three-dimensional caves and researching into cave painting, cave dwellers, cave deposits, etc. Who knows where it will lead?**

Monica Deahurst,
Albany High School, Chorley

Margaret Chlond
Aged 13
Albany High School
Chorley

Gavin Redgriff
Aged 4
Strand First School

Joanne Long
Aged 7
Pelham Infants School
Immingham

Stephen Garrett
Aged 7
Rockingham Junior and
Infant School
Rotherham

Suggested activities

1 (a) Children can be asked to consider the linear qualities to be found on the surface skins of whatever variety of animals and insects are available.
(b) These surface skins will also produce particular focal points, both in decoration and organic form. At the same time, animals and insects themselves can often be the focal point in an environmental setting: the fly on the wall, the sheep in the field. Ask children to record appropriate focal points.
(c) Observing shapes may be more practical in terms of smaller creatures such as insects.
(d) Fur, skin, fleece and scales all reveal patterns, as can collections of creatures themselves, for example, ducks in flight or sheep in a pen.
(e) The form of animals and birds is their essential ingredient, yet it gives them the very liveliness which makes them so difficult to record. Make opportunities for children to draw elements of that form.
(f) A productive idea to explore is the range of different textures to be observed in just one creature.
(g) Similarly, the range of texture, surface decoration and three-dimensional form in this area produces a good deal of tonal variation that can be observed and recorded in a series of sketches.
(h) Exploring the element of colour is again appropriate to all these activities. Look out for the way an animal, bird or insect can introduce a contrast or flash of colour.

2 A linking theme, such as 'Wings', 'Camouflage' or 'Habitats' can prove fruitful, as can something more complex such as 'Movement' or 'Flight'. Ask children to make sketches and collect photographs and illustrations on the theme, for use later in some imaginative re-creation.

3 A comparatively long-term project is to sculpt an animal or bird in clay using a model.

4 Recreate either in two-dimensional or three-dimensional form, a group of animals, birds or insects in their natural habitat, working from models, illustrations, photographs and sketches.

5 Possible themes for further imaginative work:
(a) In the cold morning, the cows waited for the herdsman.
(b) The moths settled in the light.
(c) Seagulls filled the town.
(d) A fox slipped quietly along the embankment.

A closely-observed drawing of a frog has provided the information and awareness of colour and texture to enable the artist to produce this lino print

Andrew Cole
Aged 18
Merrywood Boys School
Bristol

Characters, People and Faces

Ultimately the environment has to be seen as something created and shaped by people, so this last area of interest must be significant. The human figure too rarely appears in drawings, paintings and three-dimensional work concerned with landscape and the environment. It is perhaps significant that this and the previous area of interest are the ones which create most problems in terms of producing visually based work, probably because both figures and animals are considered difficult to draw.

Characters, people and faces give context to the environ-

Karen Dolby
Aged 10
Parlaunt Park CM School
Slough

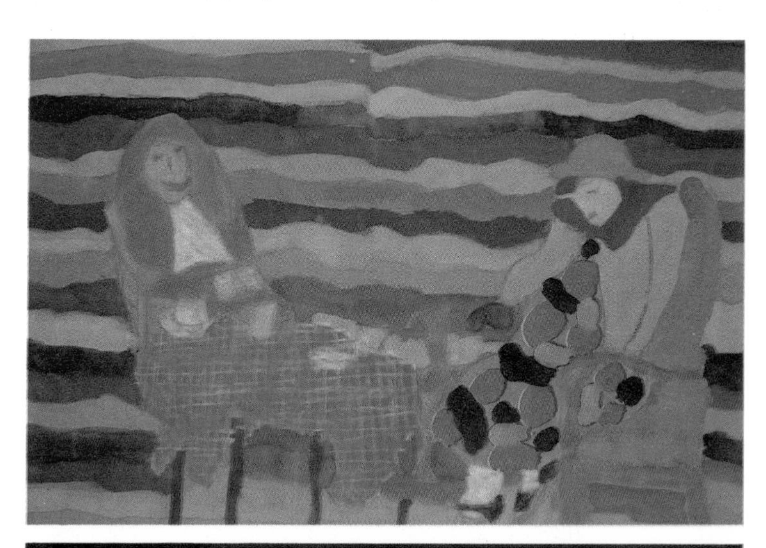

Peter Cronin
Aged 10
Parlaunt Park CM School
Slough

Pupils can dress up and be arranged in an appropriate group and drawn as models

Very young children are often able to work directly from their imaginative experience without the need to refer to models

Christine Vick
Aged 5
Nunsthorpe First School
Grimsby

ment: figures working in a street, building site or field, or highlighted against an horizon, or standing, leaning, talking or crowding an area — all create some meaning through the way they are drawn and the role they are given. Their role (what they are shown as doing and the way they are dressed) may be subsidiary to the way they have been drawn, i.e. the colours used, their shape and form, their collective pattern, the solitary use of lines.

There is a ready supply of models, from people to be observed in the environment itself — the old lady on the park bench, the man with his shopping, workers in the fields — to children in the school, colleagues or parents, dressed and posed appropriately. Again, such drawings and sketches can be used with other work to create a particular image.

The addition of the figure to an environmental image can bring depth, meaning and sensitivity to what is otherwise purely descriptive. Classic literary examples are Dylan Thomas' *The Hunchback in the Park;* and *Michael* by Wordsworth. Alternatively, a figure can add graphic description to a neutral drawing, expressing, for example, the loneliness of a housing estate, the crowded nature of a football terrace, or the remoteness of a farm worker.

This print captures
atmosphere of a q
moment in the day
the observation an
positioning of the
characters

Pat Routledge
Aged 16–18
Hereford School
Grimsby

Suggested activities

It is unlikely that children will be able to study this area of interest, in terms of collecting visual information using the outlined strategy, in the same way as the others, although it may be possible if a shopping centre, market or similar busy area can be visited. On the other hand, examples can be gleaned from magazines and newspapers, by extracting sections of photographs. Stick these onto worksheets or into sketchbooks, to use at a later stage when developing imaginative ideas.

1 (a) A queue of people, or a group sitting on a bench or huddled together on a terrace may become a 'line' in the sense that they draw across and divide a particular space.
(b) The solitary figure in the field, the swimmer in the pool, the group on the beach and the trio in the park — all these are possible focal points. Encourage the children to find a wide range of other such examples.
(c) The angle and attitude of a person's stance, a turn of the head, the way a group has formed itself — all these provide different shapes.
(d) As a figure or group or individual face moves or turns, so it will reveal something of the form which makes it what it is. Some attitudes, faces and group relationships will convey a sense of form better than others, e.g. the bent old man, the fat face, or the rugby scrum.
(e) Lines on a face, eyes peering from a group or sometimes even the way an individual moves, may convey a rhythmic pattern.
(f) Texture is a rich source of material, from observing differences in the ageing process and how weathering affects skin texture to the way the clothes vary in texture. Any one group is likely to contain a wide range of ages, colouring and style of dress.
(g) The same element of variety will yield a considerable number of tonal qualities.
(h) There is lots of scope for colour work, too. Look at the conscious application of colour to the face and the use of colour in dress. Encourage children to seek out particular effects. Observe also the unconscious effect of colour, especially in faces, in clothes and in groups.

2 Ask four or five children to dress up like their parents, in appropriate hats, coats, shoes etc., and pose them in, for example, a queue or as figures sitting in a park.

3 Provide children with small hand mirrors and ask them to look at and draw or paint their own face.

4 Using various scrap materials, including clothes, recreate figures in an environment as a full-scale model.

5 Possible themes for further imaginative work:
(a) Potato pickers.
(b) The woman in the park.
(c) Clearing the leaves.
(d) A face: tired, worn and used.
(e) Sitting in a doorway.

Part II The Elements

Point

Point simply represents the basic mark. The shape of an empty field changes when a man enters it, because that man becomes the focal point. The balance of the space in the field around the man alters with his position. A simple way to illustrate this concept is to draw a series of empty fields and in each one place a figure in a different position.

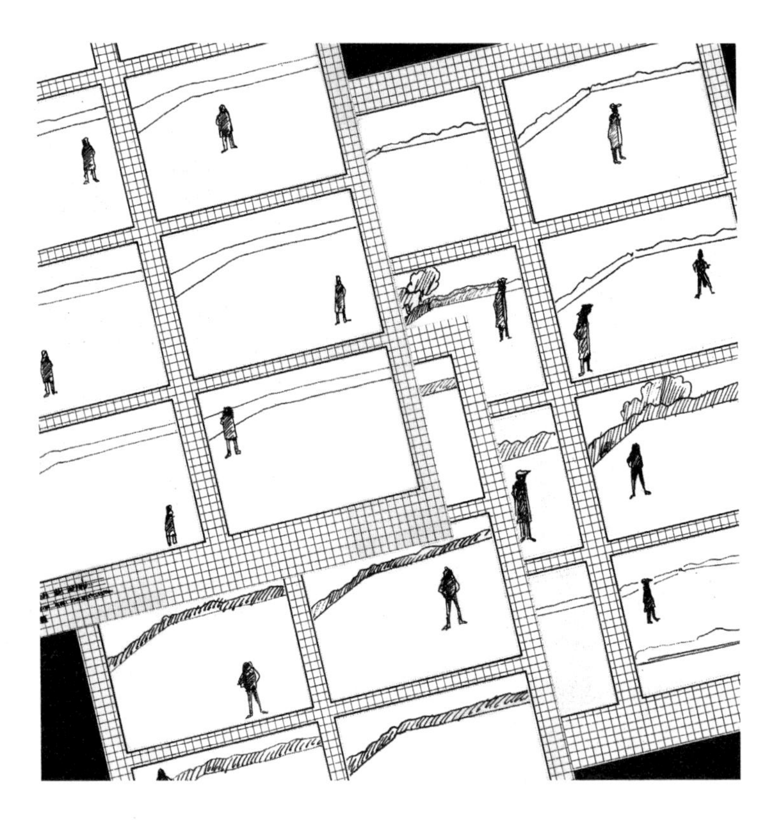

Working in an environment comprising fields or similar spaces, children can be asked to find 'points', such as figures, landmarks, machinery or trees. Equally, they can be asked to search out dominant focal points on walls or on foliage or against the skyline. This activity can be used to extend work within the classroom when children start the process of building pictures from their sketches and notes. They can be asked to place dominant 'points', such as figures, plants or animals, in their landscapes.

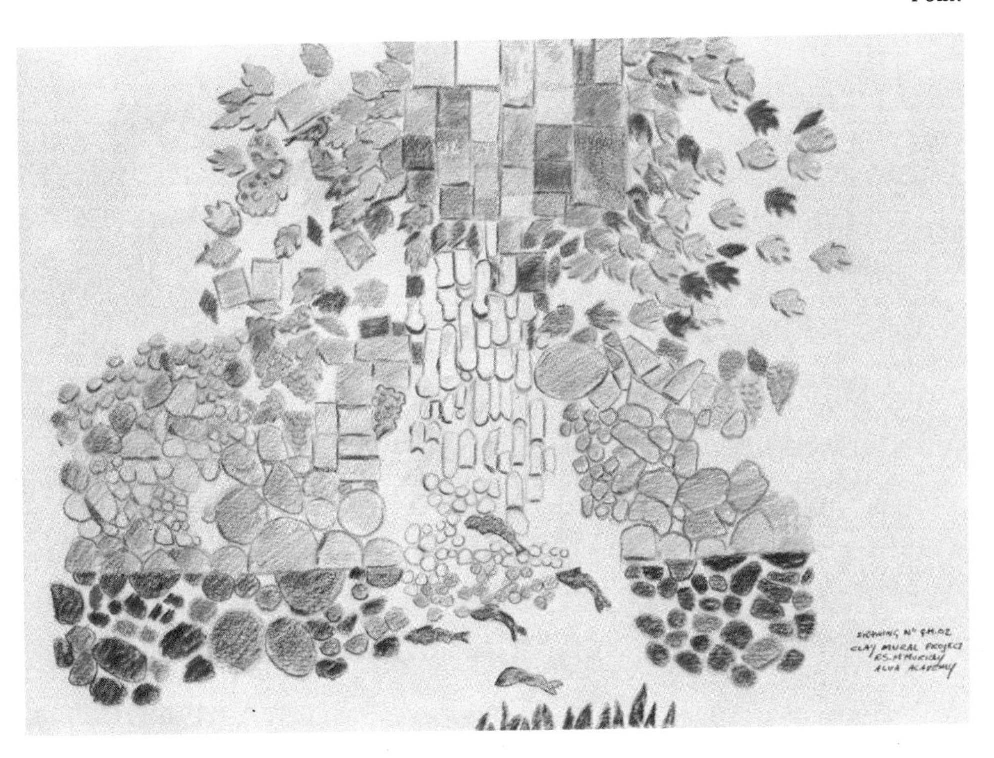

This picture can be seen as a series of focal points with emphasis given to some by the use of colour

R. S. McMuricay
Alva Academy
Alva
Scotland

The football acts as the main focal point in this picture, with the sad-faced goalkeeper and the stark lettering giving extra life to the exercise

C Drew
Southway
Comprehensive School
Plymouth

Line

Line has been described as the joining together of points, but it is better to think of it as the focal point being given a definite direction. Taking the case of the field again, replace the man by a track which follows a definite path and this time the path gives shape and balance to the field and confers some meaning to the landscape.

Equally, the line could be the top of hills against the skyline, roads through farmland, one building against another, a line of tiles against brickwork or a bridge across a river. What is important is to identify the dominant mark which draws most attention.

There were two of us and we started looking for paths. Out on the moors there were 'scars' on the surface of the landscape and I remembered Barbara Hepworth talked of roads as cutting through the landscape. She claimed that this image was used in her sculpture. . . . And into the woods, there were muddy paths, obscured partly by leaves, and bridges which paths led toward and over. The 'line' traced the direction of the path through the trees. . . . Other paths — descending, diminishing, rising — disappearing and reappearing like the cliff top path.

R. Tonks,
Headlands School, Bridlington

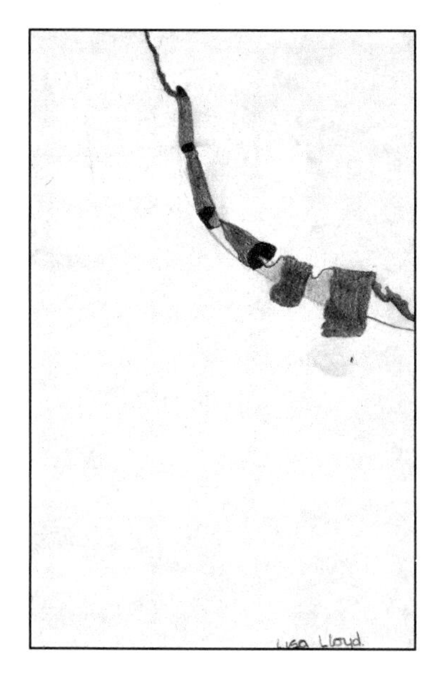

These two drawings consist largely of single lines dividing the paper, but they convey a sense of a sharp-edged hillside and undulating landscape

Lisa Lloyd
Aged 10
Claire Macfarlane
Aged 10
Parlaunt Park
County Middle School
Slough

The orderly lines created by the pine trees contrast with the direction of the carefully observed feathers of the bird, and are seen through the oblique scratches on the window pane. The lines that make up the spider's web divide the picture into an intricate pattern

K. Davies
Aged 15
Newbridge
Comprehensive School
Gwent

Again a writer reveals how seeking out the 'line' was a starting point which led to other elements and aspects of a particular environment and provided a direction for looking, as well as a purpose to his notation.

Having identified 'line' as an element, children can be encouraged to search out in detail its effect on a range of organic form and in a variety of different landscapes.

Line draws the eye in different directions and creates a sense of movement. If an environment is well known, then children can be asked to undertake particular tasks: to note, for example, the bark on certain trees, the wire by the school yard, or the scratches on the tiles.

The work on line will clearly lead on to an awareness of rhythm and indeed none of the elements should be seen as separate. It is important for children to be able to identify the different elements for themselves. This will give them a way of identifying the different elements in the visual information they are confronted with and a way of describing their response to it. In the picture by Helen Carvill on page 74, the lines are thick, strong and rhythmic. If the girl who drew it has the means to articulate what she was seeking to express, she is already further along on the journey towards greater sensitivity and awareness of the environment.

Careful observation reveals this picture to consist of a variety of lines, taking the eye in a variety of directions, creating different visual sensations by their rhythmic and contrasting use.

Helen Carvill
Aged 16
Headlands School
Bridlington

Mark Owen

Carefully cut lines highlight the texture and shape in both these crowded environments

Aged 16–18
Hereford School
Grimsby

Eric Coyle

The lines may be horizontal or vertical, angular, thin or thick, contrasting or rhythmic

James Wheeler
Aged 7
Shackleford St. Mary's First School
Godalming

Rhythm

Rhythm occurs when the line or point begins to create a sense of visual movement. This sense of movement can begin to dominate the effect of individual lines and itself become a focal point. In the picture below, the lines of the furrow take the eye in a certain direction, giving an impression of undulation and of the flow of the soil. The way these lines collectively relate gives this sense of rhythm.

Helen McCallum
Aged 11
Westwoodside CE School
Doncaster

By drawing all the lines in the same direction a sense of rhythm is created which satisfies the eye and seems to convey a sense of excitement. This 'falling rain' method of drawing can give vibrance to the most tranquil scene

Sarah Fergusson
Aged 16
Knutsford County High School
Knutsford

In the same way, lines of grass can give a feeling of breeze and movement, and can direct the eye into a particular area, just as the lines of a bird's feathers or a stone's markings can create a

The lines and shapes maintain a balanced rhythm. The shapes between the lines have become as significant as the lines. Both run in similar directions, growing or reducing in size, and a sense of rhythmic winter stillness is conveyed

Christopher Clarke
Aged 16
South Hunsley School
North Ferriby

In this picture the trees have been drawn to convey the rhythmic movement of the wind

Karen Tomkinson
Aged 12
Ryecroft Middle School
Uttoxeter

rhythmic form. But this comes not only from line. It may be created by a collection of similar shapes placed together, such as those made by branches of trees, leaves on a bush or stones in a wall. Rhythm emerges from relationships: whether relationships of line, point, shape or whatever. It is important to ask children to look for relationships that create a sense of movement.

The assumption behind the exercises and suggestions in this book is that teachers and pupils are learning, whatever the level of skill and knowledge at which they start and whatever their motivation — and one of the values in looking for these elements of visual analysis is an improvement in the ability to draw. By looking at rhythm in the environment, children may understand that if they consider the relationship of the lines and shapes they draw, then they can create effects of movement in their work.

Shape

Shape is the outline of something, but whether this is the outline of fields, buildings, trees or creatures, or whether the spaces between objects, shadows or colours, is immaterial.

It is the outline of a form or an object which gives it its strongest sense of meaning. Frequently drawings do little else but convey the shape. The value in looking at and encapsulating shapes on paper lies in providing an object that can be worked on. After the shape has been established, attention can be drawn to form, texture, pattern and colour. However, sometimes it is worth taking a longer and harder look at outlines. Just as children often think of sky as blue and sand as yellow, and fail to see their true colours, so familiar shapes become clichés, and children need to be encouraged to look again at the outline of trees, houses and a variety of animals which they know largely from cliché and caricature in children's books. Ask them to look at the unusual or unfamiliar outline, for example, space between objects, colour or shadow, as well as something familiar but perhaps undrawn, for example, local fields.

Esther Geldard
Donald MacBride
Stephen Osborne
Lindsay Allan
Sandy Bartai
David Horne
Aged 11–14
St. Mary's Music School
Edinburgh

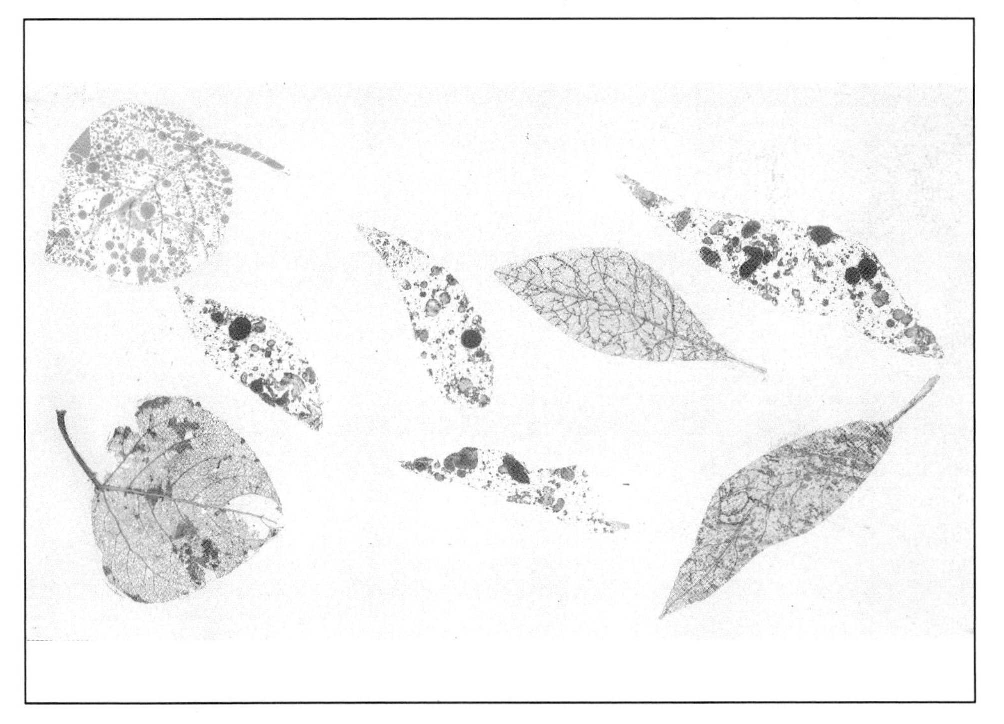

Irregular shapes, straight edged. Unexpectedly angular. Fields laid over the more ancient roll of the land — man's agricultural landscaping.

Jean Hockenhull,
Riddings Comprehensive School, Scunthorpe

But the shapes may not always be in the landscape. Familiar objects in the school, studio or home may provide a useful starting point. An organised still life may provide a means toward breaking the cliché of remembered shapes.

The smooth outline of the spilt syrup contrasts with the detailed outline of the bee.

Sian Teasdale
Aged 15
Gillott's School
Henley-on-Thames

Form

Form is shape with depth; the three-dimensional outline. It is the quality within the drawing through which real knowledge of the subject is conveyed.

Gavin's innate visual interest in nature took the form of copying from illustrations from about the age of seven. When I first noticed Gavin's talent, at the age of eleven, I felt it important that he should draw and paint from first hand experience and consequently set before him all manner of natural objects from which to work. . . . These included fish, prawns, pheasants and other game birds. His natural handling of materials also helped him to use new media, including aguanell crayons, water colours, pastels, pen and ink and etchings. . . . More recently he has ventured outside to make studies of the environment and local landscapes.

R. D. Davies,
The King Edmund School, Rochford

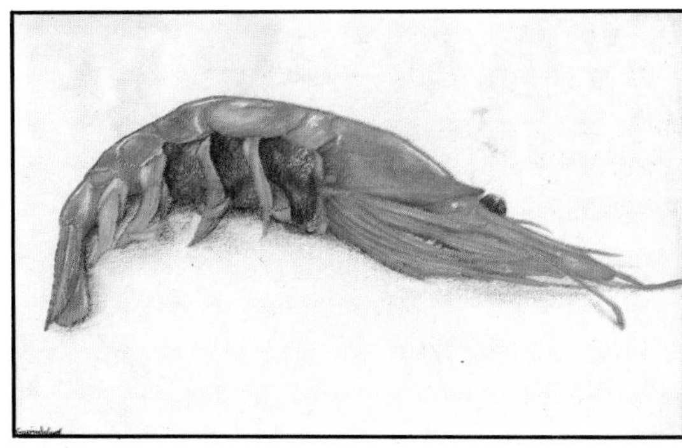

Gavin Ward
Aged 13
The King Edmund School
Rochford, Essex

Pupil aged 14
Vale of Ancholme
School
Brigg

The use of tone conveys something of the three-dimensional quality of these flowers

Michaela Leech
Sidmouth Junior High
School
Hull

It is not sufficient to just draw from photographs or illustrations, because all that is required is a particular kind of technical skill which does not encourage thought, feeling or awareness, except in a rather indulgent way. Copying can be a pleasant, relaxing pastime but not a learning activity. In order to draw an object, it is important to have some knowledge of it, some sense of its size, feeling, smell and movement.

Reference has already been made to the importance of handling the objects to be drawn. The value of making a clay model of something is not necessarily the sculptural form itself but the fact that the process requires the pupil to look at all aspects of the object. And that is 'form'; it is the total sense of the object, whether landscape, building or organic matter.

One class was investigating the qualities of form and space and had already had some lessons discussing and experimenting with this subject when I took them to examine a local hedgerow. The hedgerow contained a wide variety of flora by virtue of its being old, flanking what was once an old cartway, now just a farm track. I told the children initially to choose something that they found interesting. At first the obvious things like toadstools and berries caught the eye but gradually they began to see more.... One or two spring flowers had emerged in the warmth of the autumn sun and this became another source of discussion which kept their interest up while the hard work of analytical drawing began. They were all using a 2B pencil on white cartridge. Some became so interested in all that was around them that they added as many different specimens as they could find to their drawing. Some sorted out the pattern of the undergrowth, while others looked through the hedge to the field beyond .... When the drawings, done entirely on the spot, were completed, I asked the children to produce a picture, this time in the art room, about the lane which they now knew well. I asked them to show which qualities appealed to them most. Some children painted, some used coloured pencils, some oil pastels, some drew freely with a brush, others worked in pen and wash, while several chose to do progressive lino or screen prints.

R. D. Davies,
The King Edmund School, Rochford

Pattern

Pattern is a difficult element to grasp, perhaps because it appears to mean something more than it actually does. It refers to something which can act as a model or an example, rather than an object in its own right. In the environment this could be something which is distinctively familiar, e.g. a tree or collection of trees, particular leaves and arrangements of plants, or the shape of buildings or architectural forms.

The subject seemed very broad at first — the more one thought about it, the more possibilities emerged. It helped to tie the approach down, from the beginning, to considering one aspect at a time and, when collecting material, to bear in mind how it might ultimately be developed or treated. In this way, it should be possible to collect enough material in a day, or to use work on the actual location for a more sustained project. . . . Simply looking at pattern and open spaces without taking it along other possible avenues is an enormous topic. My understanding of pattern was an element which embraced the relationship of shapes, random or symmetrical, and repetition of form or texture within the different constituents of the landscape. . . . Any individual's approach to such a wide topic will differ according to experience and interests. Fields, for example, would seem to involve also looking at fences, walls, plants, stones, trees, hedges, cultivation and so forth. . . . My first decision was to consider the space between objects, probably because much of my own interest and experience in recent years has been in printmaking. The first drawing was very simplified and rather stylised. I found that as the time progressed, I worked toward a much freer and more spontaneous treatment of subject matter.

Jackie Goodman,
Hessle High School, Hessle

One of the reasons for the misunderstandings and confusion about pattern is the association it has with decorative form. Children are frequently asked to go out and gather 'patterns' that they can use decoratively. Thus they connect pattern with familiar or hackneyed shapes and assume it must be something someone else has identified as appropriate. They search for 'patterns' which are like other patterns.

There is a need to break this mould. One teacher has approached the theme as follows, using 'fields and spaces' as an area of interest and the element of 'pattern' as providing a direction for exploring it.

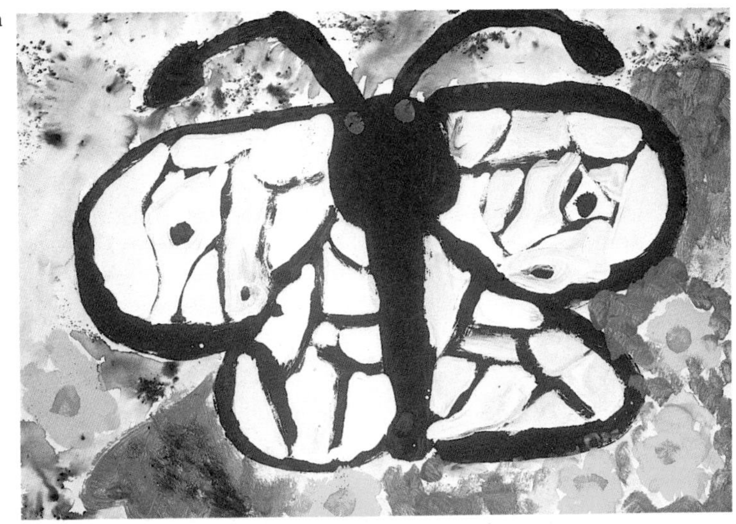

This painting works as a model or pattern in the sense that it is recognisable as an archetypal butterfly

Rachel Tidswold
Aged 8
South Parade First School
Grimsby

Observing the rhythmic placement of bricks reveals a model for repetitive pattern-making

A Hebden
Aged 14

The row of houses provides a model for repetitive pattern

Suzanne Bradley
Aged 6
South Parade First School
Grimsby

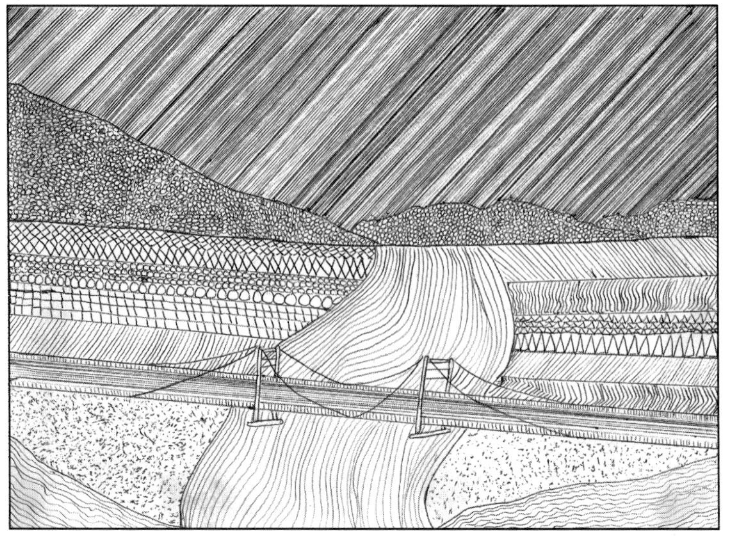

A simple landscape has been imaginatively developed into a decorative design

Heather Malton
Aged 12
Mersey Junior High School
Hull

This print is the result of observing the arrangement of trees, windows, fence and borders, which can be seen as distinctive patterns

John Barker
Aged 16
Hereford School
Grimsby

Fields and Spaces

Pattern

A project for studying pattern within the environment of the school.

1 Material: each child should have a portable, personal kit consisting of:
- **(i) Pencil + sharpener**
- **(ii) Coloured pencils**
- **(iii) Watercolours/brushes/water**
- **(iv) Charcoal**
- **(v) Water-based felt tips**
- **(vi) Oil pastels**

2 The teacher should first investigate the school area and identify particular patterns. First thoughts include:
- **(i) Bicycle wheels/motorbikes in sheds**
- **(ii) Windows in multiple — sides of buildings with windows**

(iii) Looking through regular shapes, railings, decorative walls
(iv) Areas where the visual stimulus has clearly defined areas/shapes/patterns which can be identified

The materials most useful for this are water-based felt tip (black), pencil and water colour.

3 A conversation/discussion/demonstration should take place as to what is meant by 'pattern' in a deliberate attempt to give pupils the concept of what is involved. The difference between pattern and texture should be emphasised.

4 The next stage is for children to work in the environment, drawing elements of pattern as previously identified by the teacher.

The materials most useful are pencil, black felt tip, water colour and coloured pencils.

5 Field work is important for pupils and they should be carefully briefed. They should be asked to look for shapes which repeat, either identically or progressively, for example:
(i) bicycle wheels from elipse to circle — overlapping
(ii) brick walls against hedges, against windows
(iii) shapes of doorways at junctions
(iv) shapes between bases of hedges
(v) shapes between branches and shapes of trees
(vi) shapes of sky between tree and hedges

n.b. The emphasis is on the potential pattern of shape areas.

6 The classwork should begin by looking at the drawings and helping define where progression is occurring. Attention should be drawn to relationships through extraneous subject matter which might enhance or possibly bring in a new/second set of pattern relationships.

7 School Field Project:
The area of activity is concerned with the edge of the school field, bounded by a row of concrete posts which once had a wire fence between them. Much of this has rotted, broken or been damaged and hangs from the posts. Behind this fence there is an overgrown bank with nettles, different grasses, bindweed and trees along the top.

The posts should be numbered by the teacher. Pupils should be taken out and asked to draw 'visually interesting situations' along this edge, noting the numbers within which they are looking and working. A discussion should take place to identify what is meant by the 'visually interesting'.

The approach might be through rhythm. The definitions of rhythm:
(i) Man-made against natural form.
(ii) Wire mesh against a concrete post with grass strands.
(iii) Shapes of trees against skyline/sky.
(iv) Old growth against new.
(v) Rubbish against natural form.
(vi) Leaves against tree.

E. Ullyot,
Andrew Marvel School, Hull

Colour

Colour may be the most obvious of the elements given for investigation, but it is also one of the most difficult, not least because it is a minefield of accepted clichés.

The school is situated in the Kentish Weald and is lucky enough to be surrounded by farmland. The majority of pupils live in country areas but sadly I have found that often these are the ones least aware or appreciative of their environment. In the age group that I teach, 9 to 13, one of the most important roles of the art lesson is to develop the children's ability to look; so that they see not what they expect to see but learn to look and discover what is there. I find the easiest way to explain these two ways of looking to the children is to ask, for example, the colour of the sand by the sea. More often than not, the answer will be 'Yellow'. On further questioning such colours as beige and cream may be offered. I might even show them some examples of different coloured sand alongside a piece of yellow paper so that they can see how unsuitable yellow was as an answer. I will then ask them why yellow was suggested. Gradually, with continual questioning, they work

Sketches, notes and photographs were made during a visit and used to re-create the colours found in a specific area of the environment.

Juliet Walker
Aged 11–14
Sheffield High School
Sheffield

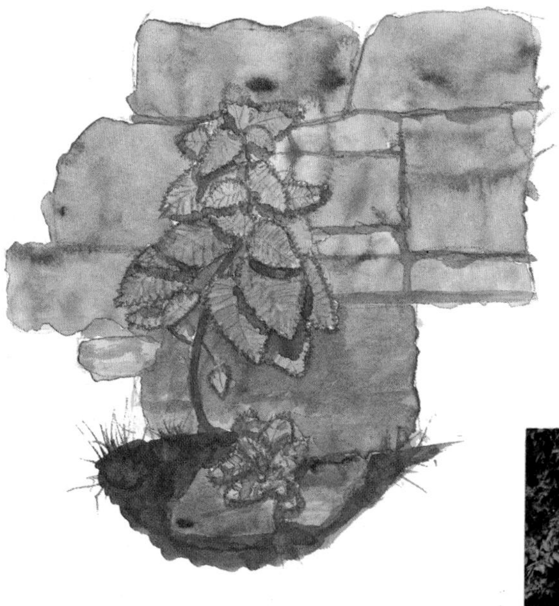

out that it is because early learning books depict sand in a bold yellow, beige being too subtle a colour for a toddler to appreciate or learn to say, and that this has become imprinted on their minds to such an extent that they actually see yellow.

L. A. Dear,
Dulwich College Prep School, Kent

There is no better way of overcoming this problem than by asking children to go out and note the colours that they see. The difficulty is that few of them possess a language sufficiently rich to communicate the range of hues and tones that confronts them. However, they will not become aware of colours by having them

The rich, vibrant quality of this picture is the result of picking out the bright, warm colours found in the environment

Samantha Sumpter
Aged 12
Willows Middle School
Grimsby

This picture achieves its effect by concentrating on the harmony of autumn colours

Dawn
Aged 8
Holme Valley Primary School
Scunthorpe

shown and described, only by looking and by being encouraged to see for themselves.

Areas, spaces or objects with specific but limited colour will need to be identified and visual aids such as photographs and small-scale sketch plans will help children find an area within which to work. The child will also need a carefully selected palette of colour, whether water colours, crayons or pastels.

It is important to recognise the difficulties facing the child who is asked to record or even recreate some aspect of the environment in colour. Not only must they reproduce the outline, form, texture, line and tone correctly, they must reflect the nuances of colour — all too frequently with inadequate paints, poor quality crayons or even the 'wrong' colours. They may be given a mass of colours to use — inappropriate but so tempting! It is hardly surprising that colour often becomes a matter of 'filling in the black and white' — not helped by the way graphic artists often use caricatured colourisation.

The pupils were shown slides and actual examples of brightly coloured birds by the school caretaker who is a breeder. As a result of this they chose the species and the colours (a maximum of three) which they wanted to use. It was interesting to see how, with a restricted colour range, they coped with a full colour ranged subject, mostly by the use of textural cutting.

P. Needles,
Merrywood Boys School, Bristol

Tone

Tone is the variation in light that is reflected from an object, although it also refers to the range of variations possible within one colour. It is that aspect of drawing frequently referred to (by children and some staff) as 'shading' and therefore too often seen as a technique to be acquired rather than the result of careful observation.

The use of a lamp or spotlight in the classroom to highlight a pile of pebbles is an easy way to draw attention to tone. Similarly, the lamp can be used for other still life arrangements, including the stuffed animals referred to earlier. It may even be practical to build small natural environments within the classroom and highlight them in this way; drawing attention to the way the fall of light subtly changes the shape and the form. It is often difficult to see this in the environment, simply because of the mass of information that constitutes the landscape.

Tonal exercises should involve colour. They should not remain mere explorations of the gradation possible with a pencil or graphite stick. In the first instance, we live in a world of colour and, in the second, colour can be a more exciting area to work within.

Careful observation of the tonal differences in this corner of a field helps reveal the delicacy of shape, variety of form and fine detail in the environment

Stuart Pringle
Aged 12
Dulwich College Prep. School
Kent

It is worth noting places around the school and in the environment where the exercise of noting tone can easily be undertaken. A corner of a field, a view through a window or across parkland may be identified which allows the child the opportunity for a personal statement while providing useful limitations on the information to be gathered.

The pupil has concentrated on the change and contrast of tone in this picture. The gradual change in tone is created by the use of rhythmic lines.

Elaine Young
Aged 14
Henry Cavendish School
Derby

Sonja Titley
Aged 11
Dulwich College Prep. School
Kent

Stuart Barlow
Aged 14
Albany High School
Chorley
Lancs

Texture

Collage, carefully used, can be an excellent way of reproducing textures observed in an environment

Emma Fisher
Aged 14
St Gabriels School
Newbury

Texture refers to the representation of a surface, which could be anything from the now familiar field, to the broken stone of a limestone wall, the mottled shell of a dead snail, or the fine tissue of a dried leaf. All objects, whether man-made or natural, have texture and it is important to direct children's attention to the various qualities of the differing textures.

Individuals do not perceive the same object, environment or element in exactly the same way. Twenty children will draw one piece of bark in different styles, accentuating differing features. Clearly varying degrees of skill influence the final result, but the main reason why drawings differ is that what we see is filtered through our own experience. Obvious examples are the child, once cut by a stone, who accentuates its jagged edges, or the child, fearing an attack by a large bird, who stresses the sharp beak and predatory look of the owl. And in our teaching we should be equipping children with the appropriate 'language' to do this, so that they can organise and make sense of their experiences. We should be encouraging children to use their skills to create a picture of the environment as they see it, not as they think we want them to see it. We need to help them refine and develop their ability to achieve this personal vision.

The examples below show how children asked to look for and respond to textures in a wood can produce some surprisingly

Nicola Birch
Aged 11–14
Headlands School
Bridlington
North Humberside

powerful and individual results. They reveal a dramatic, almost emotional response which doubtless came from the dialogue initiated by the teacher. In contrast, the following report exhibits a more controlled yet thoughtful approach:

> **I walked down the cliff road to the bay, first looking for interesting surfaces on the tree trunks. There were several whorled and knotted areas of bark which I photographed and made sketches from. The photographs were a means to remind me of the experience — photographs so rarely work without the original experience — and to use later in classroom work. . . . It seemed that the obvious way to tackle the problems of this project was to collect flowers, leaves and bark and take them back to the studio. They could be drawn there, rubbings and**

Simply concentrating on texture, as in this picture, may be an effective way to depict the subject matter.

Jamie Lishman
Aged 7–10
Hambledon Primary School
Henley-on-Thames

Techniques using wax-resist, stains or a combination of media can be most appropriate in seeking to re-create texture

Gareth Stringer
Aged 13
Sir Hugh Owen School
Caernarfon

Brickwork, tiles and shuttering are the texture here

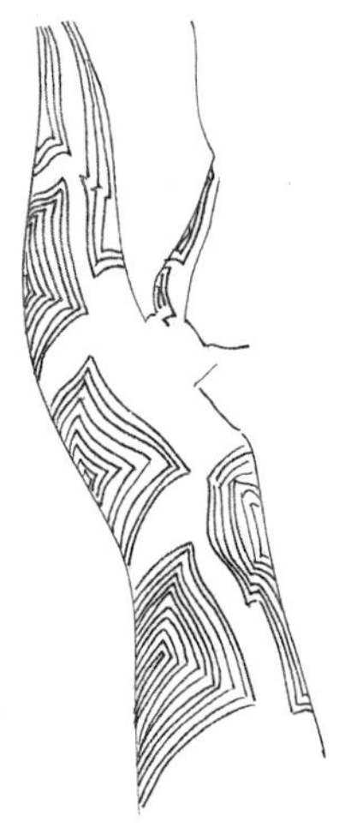

Shape/pattern might be extended from this sketch

casts made and they could be generally studied — but having worked this way so often, it was difficult to muster enthusiasm. The nearer I got to the beach — seeing that endless stretch of sand, the arrangement of stones stretching across the mudflats — the more I wished I had not been limited to looking at the fauna. . . . The textural qualities and patterns of stones and cliff face seemed so much more exciting than the area with which I was concerned. I could work up no enthusiasm for it and so decided to ignore the brief and seek my own inspiration and motivation. . . . I wandered through a field gradually finding my interest drawn toward the variety of trees and then I saw a single tree, standing solitary in a field. The isolation of the tree impressed itself on me and I sat and drew it over and over again. I had never been able to draw trees; no art school training to help me with the skills and techniques I felt necessary to reproduce what I could see in front of me. . . . I struggled with my pencil to capture the different qualities of tree and branch and in moments of that struggle I suppose I achieved that loss of consciousness when one is at total empathy with and for that which is being drawn.

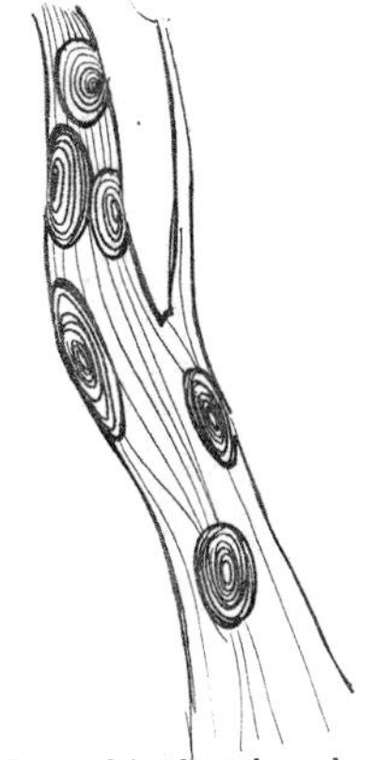

From this sketch, colour studies could be developed

Trees, flowers, shrubs and grass: Texture

1 **The aim is to find an area of personal interest — many of the above will have interesting surface qualities but some will have more immediate appeal. The objects may be large or small and the area may be an isolated part of a highly complex whole.**
2 **The next stage is to make a drawing of the selected 'area' — the tree, flower, shrub or grass. In this instance, the object drawn was a section of a tree. The immediate appeal was the texture of the bark — very broken and coarse — very craggy in contrast to a smoothness elsewhere on the tree and surrounding shrubs. Two sections of the tree were drawn. Further drawings were made from the original sketches. (*Note:* The initial sketch of the tree was very brief. I continued to look for other items of interest and, in fact, found knotted sections of shrub, highly linear, and later a red berried bush but it was the texture on the tree which had the strongest appeal to me. I returned to it and spent a longer time investigating the tree. It subsequently formed the basis for my interpretations of the project).**

G. Gilbride,
Driffield School
Humberside

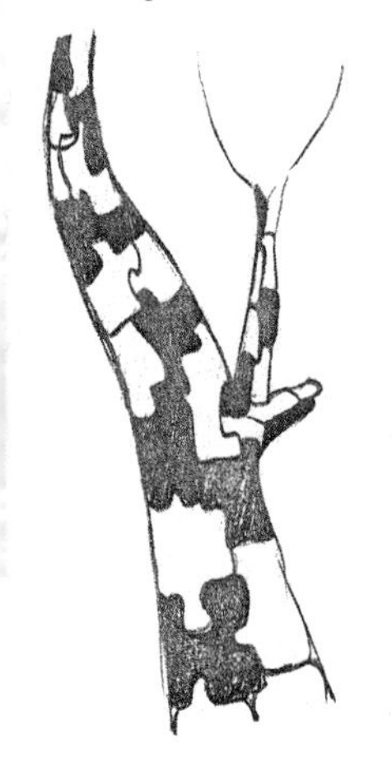

A jigsaw puzzle — which could be developed as:
(a) Collage — incorporating alternative, found, man-made/natural textures
(b) Each jigsaw section could contain an alternative area of the landscape

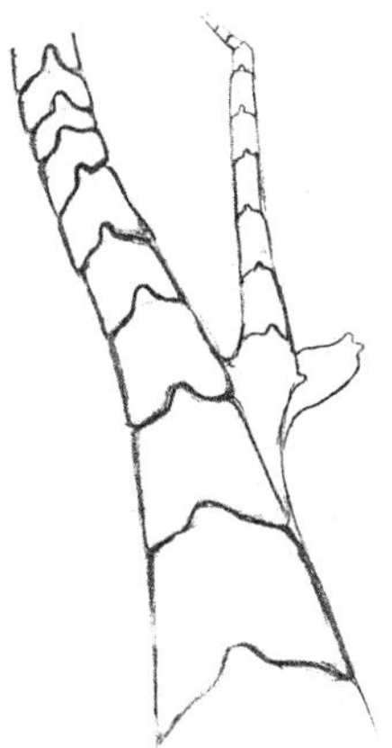

An attempt to make the tree look as if it is made from a totally different material, e.g. rope, string, layered leaves

Conclusion

It has been emphasised throughout this book that the ultimate aim is to encourage children 'to observe the fine detail of the small place', and to learn to know through seeing. The object is to help them to understand better the world around them and, through that experience, to understand better the world within them.

I believe that in studying nature, one comes into contact with a universal touchstone that demonstrates an underlying order in its elemental forms, so enabling the individual to, in some way, come to terms with the two worlds which exist: that which exists outside himself, whether he exists or not, and that which exists only for him, in his imagination and personality. To quote Bronowski . . . 'The notion of discovering an underlying order in matter is man's basic concept for exploring nature. The architecture of things reveals a structure below the surface, a hidden grain . . .'.

J. Teasdale,
Knutsford High School, Cheshire

Thus the ideas expressed within this book are concerned as much with helping pupils discover qualities within themselves as with understanding the environment. One of the main functions of art within the curriculum is to help children make sense of their emotional, imaginative and sensory experience. Much of this experience will have been shaped by their physical surroundings:

This was the London of my childhood, of my moods and awakenings: memories of Lambeth in the spring; of trivial incidents and things; of riding with Mother on top of a horse-bus trying to touch passing lilac-trees — of the many coloured bus tickets, orange, blue, pink and green, that bestrewed the pavement where the trams and buses stopped — of rubicund flowergirls at the corner of Westminster Bridge, making gay boutonnieres, their adroit fingers manipulating tinsel and quivering fern — of the humid odour of freshly watered roses that affected me with a vague sadness — of melancholy Sundays and pale-faced parents and their children escorting toy windmills and coloured balloons over Westminster Bridge; and the maternal penny steamers that softly lowered their funnels as they glided under it. From such trivia I believe my soul was born.

Charles Chaplin
My Autobiography, Penguin

As Chaplin reflects on his past, he recalls the physical detail which evokes sensations and feelings of childhood within him. It is important children are given a language which will help them to make sense of the physical world within which 'the inner landscape of their mind' is formed. If they can be made aware of their surroundings, they will have an opportunity to understand its possible influence upon them. The visual arts can provide them with a skill which enables them to recognise and explore the environment, and a language which will help them to develop their experience of that environment. Indeed, the artistic process involves skills which can be used in a variety of information – gathering activities. These skills are likely to become permanent tools for the process of learning, remaining part of a person's vocabulary for the rest of their life.

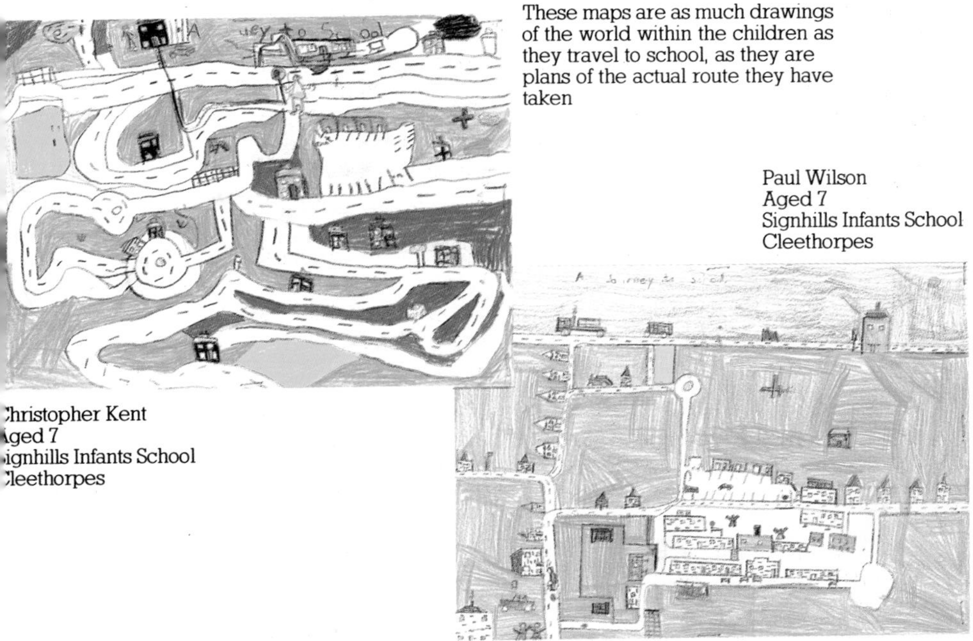

These maps are as much drawings of the world within the children as they travel to school, as they are plans of the actual route they have taken

Paul Wilson
Aged 7
Signhills Infants School
Cleethorpes

hristopher Kent
ged 7
ignhills Infants School
leethorpes

Bibliography

1 Blythe, R., *Akenfield: Portrait of an English Village*, Penguin Books Limited
2 HMI, *Art in Primary Education*, HMSO, 1979
3 HMI, *Art in Secondary Education, 11–16*, HMSO, 1983
4 North East Art Advisers, *Learning Through Drawing*, NEAA, 1979
5 Read, H., *The Innocent Eye*, Faber & Faber Ltd
6 Read, H., *Education Through Art*, Faber & Faber Ltd
7 Stanislavsky C. (Trans. E. R. Hapgood), *Creating a Role*, Eyre Methuen
8 Witkin, R., *The Intelligence of Feeling*, Heinemann Educational Books

Glossary

Artistic process 52 The process of turning an imaginative, sensory or emotional experience into an art form.
Batik Method of dyeing textiles in which each dye is applied separately while the cloth not to be coloured is covered by wax.
Black conte crayon Crayon which gives thick, strong marks and which is reasonably fast.
Collage The use of a variety of materials to create a picture or form.
Drawing sheets/worksheets 12, 21, 70 Sheets with organised spaces into which students can gather their visual notes and sketches.
Environmental furniture Everyday objects that are regularly found. For example, in streets these would include telephone boxes, signs and bus-stops.
Etching Method of printmaking based upon drawing into a resistant surface.
Fabric work Includes collage, batik, needlework and printing on fabric.
Focal point Mark which draws attention; the point on which the eye focuses.
Graphite stick Thick black crayon-like stick, similar to charcoal but capable of making a much stronger and more permanent mark.
Ink wash The use of drawing ink, usually thinned with water, to lightly cover a surface, often one which has a picture or drawing done in waterproof ink or wax crayon.
Machine embroidery The use of a sewing machine to produce the effect of embroidered line, shape and/or texture.
Mono-printing Medium which produces only one print.
Narrative interpretation The use of art to express or imply a story or dramatic event.
Objective drawing Drawing from life in a direct attempt to capture what is seen.
Oil-pastel Pastel bound together with an oil base which gives a rich, permanent colour.
Organic form Natural form and shape as opposed to man-made but implying growth, for example plants, roots and bones.
Over-laying wax colours Wax crayons used on top of each other to increase richness of colour and variation of texture. Oil-pastels can be similarly used.
Pen and wash Combination of pen and ink together with brush and water.
Printmaking Methods of transfering a mark from one surface to another, ie lino, wood, metal or card to paper or fabric.
Resources/visual stimulus collection Any collection of artwork or artefacts maintained by a school/teacher/department as references to stimulate work and ideas.
Screen printing Method of printmaking in which ink is pressed through a mesh-covered screen on which a stencil has been mounted.
Sculptural form Form which is created in space to convey a specific artistic concept.
Sepia conte pencil Wood pencil crayon which creates a rich brown colour.
Skying The process of making drawings of the sky, as in the sketches of John Constable.
Subjective drawing Drawing which emphasises the artist's personal response, interpretation or reaction.
Three-dimensional work Work which has depth in space as well as height and width.
Tonal qualities The quality of light and darkness; the general effect of light and shade.
Two-dimensional media The various media associated with art which have no physical depth, such as drawing, painting, printmaking and photography.
Visual analysis Analysis based upon observation, presented in a visual form using two- and three-dimensional media.
Visual record/diary A diary or similar record of events and time based on visual images rather than the written word.
Water-based felt tip Felt tip pens which have water-based ink rather than oil-based and which are washable.
Water colour Water paint made up of pigment mixed with gum and diluted with water.
Wax crayon Crayon in which the pigment is contained within wax.

Index

Reference is made in the text to certain areas of the curriculum, e.g. Natural History. These are indexed to help teachers whose curriculum organization is based upon an integrated approach. Also indexed are relevant information and activities which could be incorporated into certain topics and projects e.g. Transport, Animals.